CIVIL AIRCRAFT ELECTRICAL POWER SYSTEM SAFETY ASSESSMENT

CIVIL AIRCRAFT ELECTRICAL POWER SYSTEM SAFETY ASSESSMENT

Issues and Practices

PENG WANG
Civil Aviation University of China, Tianjin, China

Butterworth-Heinemann
An imprint of Elsevier

Butterworth-Heinemann is an imprint of Elsevier
The Boulevard, Langford Lane, Kidlington, Oxford OX5 1GB, United Kingdom
50 Hampshire Street, 5th Floor, Cambridge, MA 02139, United States

British Library Cataloguing-in-Publication Data
A catalogue record for this book is available from the British Library

Library of Congress Cataloging-in-Publication Data
A catalog record for this book is available from the Library of Congress

ISBN: 978-0-08-100721-1

For Information on all Butterworth-Heinemann publications
visit our website at https://www.elsevier.com/books-and-journals

Working together
to grow libraries in
developing countries

www.elsevier.com • www.bookaid.org

Publisher: Matthew Deans
Acquisition Editor: Carrie Bolger
Editorial Project Manager: Carrie Bolger
Production Project Manager: Anusha Sambamoorthy
Cover Designer: Matthew Limbert

Typeset by MPS Limited, Chennai, India

CONTENTS

ABOUT THE AUTHOR

Peng Wang is Deputy Director of the Civil Aircraft Airworthiness Certification Technology and Management Research Center (CAAC) and Associate Professor of Civil Aviation at the University of China. He holds a master's degree in safety management from ENSICA, France.

He has 12 years of experience in system safety analysis, airborne electronic hardware engineering research and certification, and he has involved and participated in many Chinese-type certification and abroad-type validation projects. He has also led several aviation safety-related research projects funded by CAAC and MIT (Ministry of Industry and Information Technology of China).

FOREWORD

Since the Wright brothers took the aircraft "Flyer 1" into flight for the first time in 1903, the global aviation industry has enjoyed its development for more than 100 years. While safety acts as the lifeline of the civil aviation industry and premise of civil aircraft products to be put on the market, the safety assessment is a "measuring scale" for the safety of civil aircraft products. With respect to new technologies of aviation products, the safety assessment methods have a direct impact on the civil aircraft safety and innovation.

Over the past 30 years, China's civil aircraft manufacturing industry has made considerable development and progress with the emergence of a number of new civil aircraft types such as Y-12, MA60, ARJ21, MA700, Dragon 600, and C919, keeping up with the international trend of civil aircraft industry. Chinese aeronautical industries have also accumulated certain experience during a large amount of type developments and made explorations and innovations in solving the problems related to safety assessment methods of new technologies for some aviation products based on the national conditions.

Civil aircraft airworthiness certification technology and management research center of Civil Aviation Administration of China has been committed to the research and practice of civil aircraft safety assessment. In this book, through the safety assessment of the electrical power system of a certain type of transport category aircraft, the author, in the combination with civil aircraft safety standards and practical industrial experience, demonstrates the process and methods of the safety assessment to indicate the understanding and promotion to the development of international safety standards by the aviation industry of China. In addition, it discusses how to address the issues (e.g., single-event effects) and methods (e.g., the formal methods) in current safety process.

This book provides help and reference for engineers, students, and newcomers who are engaged in airborne system safety assessment.

Bai Jie

Vice President of Civil Aviation University of China, Tianjin, P.R. China

PREFACE

This book supplements the content of the advisory material to the regulation as well as the main supporting industry standards, and tries to discuss how to efficiently organize and manage safety activities. This book emphasizes on practices and guidelines by demonstrating the contiguous safety assessment process of civil airborne electrical power system. In addition, it discusses how to address the issues (e.g., single-event effects) and methods (e.g., the formal methods) in current safety assessment process.

Chapter 1, Airworthiness Regulations and Safety Requirements, and Chapter 2, Safety Management, summarize airworthiness regulations and safety requirements and also the management of safety activities.

The main safety assessment process and safety analysis methods, such as AFHA, SFHA, PSSA, CCA, FMEA, and SSA, are discussed in Chapter 3, Aircraft Functional Hazard Assessment, Chapter 4, System Functional Hazard Assessment, Chapter 5, Preliminary System Safety Assessment, Chapter 6, Common Cause Analysis, Chapter 7, Failure Modes and Effects Analysis, and Chapter 8, System Safety Assessment. From Chapter 4, System Functional Hazard Assessment, Chapter 5, Preliminary System Safety Assessment, Chapter 6, Common Cause Analysis, Chapter 7, Failure Modes and Effects Analysis with Summary, and Chapter 8, System Safety Assessment, each chapter first summarizes inputs, detailed process, and outputs of safety assessment process and safety analysis methods, and then a case study of civil airborne electrical power system is performed. The case study, which provides the continuity through Chapter 4, System Functional Hazard Assessment, Chapter 5, Preliminary System Safety Assessment, Chapter 6, Common Cause Analysis, Chapter 7, Failure Modes and Effects Analysis with Summary, and Chapter 8, System Safety Assessment, can show the reader how to bring the whole safety assessment activities together in a logical and efficient manner.

Chapter 9, Single Event Effects in Avionics, discusses how to address the issues of single-event effects in safety assessment process. The sensitivity assessment method is included, as well as an example of single-event effects, sensitivity assessment of electrical power system.

Chapter 10, Formal Model Based Safety Analysis Methods and the Application, discusses formal methods used in civil airborne system safety assessment, and a case study of the safety assessment for civil airborne electrical power system using the formal methods is given.

ACKNOWLEDGMENTS

I would like to thank all the people who have contributed to this book, and I especially acknowledge the following main contributions of my colleagues:

Xiao Nyue (Chapters 1, 2, and 6)

Yan Fang (Chapter 2)

Ma Zan (Chapters 5 and 8)

Dong Lei (Chapters 3, 4, and 10)

Dang Xiangjun (Chapter 7)

Zhao Changxiao (Chapter 10)

Xue QianNan (Chapter 9)

A special thanks to Cheng Wei, who provided great support and help for this book.

I would like to thank Guo Qiang, Zheng Jian, and all the other people who had contributed to this book.

I am grateful to my wife and my daughters for their years of understanding and support.

I am also grateful to Carrie Bolger, Anusha Sambamoorthy, and all the teams at Elsevier for their professional assistance.

ABBREVIATIONS AND ACRONYMS

AC	Advisory Circular
AC	Alternative Current
ADCU	Automatic Deploy Control Unit
AFHA	Aircraft Functional Hazard Assessment
AFM	Aircraft Flight Manual
AGCU	Auxiliary Generator Control Unit
APU	Auxiliary Power Unit
ARP	Aerospace Recommended Practice (SAE)
ASA	Aircraft Safety Assessment
ATA	Air Transport Association of America
ATC	Air Traffic Control
BIT	Built-In Test
BPCU	Bus Power Control Unit
CAAC	Civil aviation administration of China
CAS	Crew Alerting System
CCA	Common Cause Analysis
CCAR	China Civil Aviation Regulations
CCMR	Candidate Certification Maintenance Requirements
CI	Configuration Index
CMA	Common Mode Analysis
CMCC	Certification Maintenance Coordination Committee
CMR	Certification Maintenance Requirements
CMS	Central Maintenance System
COP	Cockpit Overhead Panel
COTS	Commercial-Off-The-Shelf
CP	Certification Plan
CS	Certification Specification
DAL	Development Assurance Level
DC	Direct Current
DCU	Data Concentrator Unit
DCTR	DC Tie Relay
DD	Dependent Diagram
EASA	European Aviation Safety Agency
EICAS	Engine Indication and Crew Alerting System
EMI	Electromagnetic Interference
EPS	Electrical Power System
ETSO	European Technical Standard Order
ETOPS	Extended Operations
EWIS	Electrical wiring interconnection systems
FAA	Federal Aviation Administration
FAR	Federal Aviation Regulations
FC	Failure Condition
FDAL	Function Development Assurance Level

FFS	Functional Failure Set
FH	Flight Hour
FHA	Functional Hazard Assessment
FMEA	Failure Modes and Effects Analysis
FMES	Failure Modes and Effects Summary
FTA	Fault Tree Analysis
GCU	Generator Control Unit
HAS	Hardware Accomplishment Summary
HCI	Hardware Configuration Index
HIRF	High Intensity Radiated Fields
ICAO	International Civil Aviation Organization
IDAL	Item development assurance level
IDG	Integrated Drive Generator
ILS	Instrument landing system
IMA	Integrated Modular Avionic
INV	Inverter
LACTR	Left AC Tie Relay
LGCU	Left Generator Control Unit
LPDA	Left Power Distribution Assembly
LRU	Line Replaceable Unit
MA	Markov Analysis
MBSA	Model Based Safety Analysis
MEL	Minimum Equipment List
MFD	Multifunction Display
MMEL	Master Minimum Equipment List
MOPS	Minimum Operational Performance Specifications
MRB	Maintenance Review Board
MRBR	Maintenance Review Board Report
MSG	Maintenance Steering Group
NPRD	Nonelectronic parts reliability data
OEM	Original Equipment Manufacturers
PASA	Preliminary Aircraft Safety Assessment
PDA	Power Distribution Assembly
PHAC	Plan for Hardware Aspects of Certification
PLD	Programmable Logic Devices
PRA	Particular Risk Analysis
PSAC	Plan for Software Aspects of Certification
PSSA	Preliminary System Safety Assessment
PTS	Purchaser Technical Specification
RACTR	Right AC Tie Relay
RAT	Ram Air Turbine
RGCU	Right Generator Control Unit
RPDA	Right Power Distribution Assembly
RTCA	(Previously) Radio Technical Commission for Aeronautics
SAE	Society of Automotive Engineers, Inc.
SAS	Software Accomplishment Summary
SCI	Software Configuration Index

SDD	System Design Description
SEE	Single Event Effects
SEU	Single Event Upset
SFHA	System Functional Hazard Assessment
SID	System Interface Description
SSA	System Safety Assessment
STC	Supplemental Type Certification
TC	Type Certification
TCAS	Traffic Collision Avoidance System
TRU	Transformer Rectifier Unit
TRUC	Transformer Rectifier Unit Contactor
TSO	Technical Standard Order
ZSA	Zonal Safety Analysis

CHAPTER 1

Airworthiness Regulations and Safety Requirements

Contents

1.1 AIRWORTHINESS STANDARDS OF TRANSPORT CATEGORY AIRPLANES

1.1.1 Airworthiness

The definition of "Airworthy" in the Oxford Dictionary is "(of aircraft) safe to fly." "Airworthiness" is the noun form of "Airworthy."

The definition of the word "Airworthy" was never included in the Code of Federal Regulations until the Federal Aviation Regulations (FAR) Part 3, General Requirements, was established. The definition was included in the guidance materials, such as Advisory Circulars (ACs) and Orders, but never in the Rule. In Part 3, the definition of Airworthy is defined that the aircraft conforms to its type design and is in a condition of safe flight.

Civil Aircraft Electrical Power System Safety Assessment
DOI: http://dx.doi.org/10.1016/B978-0-08-100721-1.00001-7

In Canadian Aviation Regulations, CAR 101.01, Subpart 1—Interpretation, "airworthy", with respect to an aeronautical product, means a fit and safe state for flight and in conformity with its type design.

Airworthiness is the measure of an aircraft's suitability for safe flight. The application of airworthiness defines the condition of an aircraft and acts as the basis for judging its fitness to fly in that it has been designed with engineering rigor, that it has been properly constructed and maintained, and that it is expected to be operated according to approved standards and limitations, by competent and approved individuals, who are acting as members of a certification organization.

1.1.2 Airworthiness Standards

Airworthiness standards are special technical standards and minimum safety standards established to ensure the implementation of civil aircraft airworthiness. Unlike other standards, civil aircraft airworthiness standards are part of national regulations and require strict enforcement.

The establishment of an airworthiness standard has involved continuous revision by accumulating long-term experience, drawing lessons from flight accidents, conducting necessary demonstration or argumentation, and soliciting opinions from the public. Thus far, FAR of Federal Aviation Administration (FAA) and Certification Specification (CS) of European Aviation Safety Agency (EASA) dominate the worldwide airworthiness standards. Taking the international characteristics of civil aviation airworthiness standards into account, many countries have established their airworthiness standards by recognizing FAR and CS with consideration of their own national situations. For instance, certification authorities from China, Canada, and Russia all have established their own airworthiness standards based on FAR and CS.

FAA airworthiness standards are as follows:
- Part-23 Airworthiness Standards: Normal, Utility, Acrobatic, and Commuter Category Airplanes.
- Part-25 Airworthiness Standards: Transport Category Airplanes.
- Part-26 Continued Airworthiness and Safety Improvements for Transport Category Airplanes. (Note: EASA also contains CS-26, but its name is Additional airworthiness specifications for operations, its initial release was on December 5, 2015.)

- Part-27 Airworthiness Standards: Normal Category Rotorcraft.
- Part-29 Airworthiness Standards: Transport Category Rotorcraft.
- Part-31 Airworthiness Standards: Manned Free Balloons.
- Part-33 Airworthiness Standards: Aircraft Engines.
- Part-34 Fuel Venting and Exhaust Emission Requirements for Turbine Engine Powered Airplanes.
- Part-35 Airworthiness Standards: Propellers.
- Part-36 Noise Standards: Aircraft Type and Airworthiness Certification.
- Part-39 Airworthiness Directives.

The above airworthiness standards are all appropriate for civil products. For materials, parts, and appliances used on civil products, they are referred to as TSO authorization according to the Technical Standard Order (TSO) standard. A TSO is a minimum performance standard issued by the certification authority for specified materials, parts, processes, and appliances used on civil products.

As of May 2016, there are 162 current effective FAA TSOs and 144 current effective EASA European Technical Standard Order (ETSOs). In recent years, despite of some slight differences in marking and data requirements, the contents of TSOs issued by the FAA and ETSOs have been basically identical in technical aspects (except the ETSO 2C series). FAA and EASA introduce mature technical standards in industry as the key technical requirements for TSO.

Airworthiness Directives (ADs) are legally enforceable rules that are applied to the following products: aircrafts, aircraft engines, propellers, and appliances. In FAA, each AD is a part of Part-39, but they are not codified in the annual edition. The authority issues an AD addressing a product when it finds that: (1) an unsafe condition exists in the product; and (2) the condition is likely to exist or develop in other products with the same type of design. Anyone who operates a product that does not meet the requirements of an applicable AD is in violation of Part-39.

AC is a recommended and interpretative material of the compliance means with the applicable regulations. Though it is stated in almost all ACs that the means it introduces are not mandatory or are not the only means, and that the applicants can adopt other methods to demonstrate their compliance with the regulation. In general, if the type certificate applicants do not propose more appropriate means, priority should be given to using the means introduced in the AC to demonstrate their compliance with applicable regulations.

1.2 TERMS AND DEFINITIONS

1.2.1 Risk

The IEC61508 Standard defines risk as the combination of the possibility and severity of hazards. It is common to use the following formulation to describe the above definition:

$$R = S \times P$$

R is the risk, S is the severity resulted from the hazard, and P is the possibility of the result.

The definition of "Risk" in Society of Automotive Engineers (SAE) Aerospace Recommended Practice (ARP) 4761 is the frequency (probability) of occurrence and the associated level of hazard [1].

1.2.2 Safety

The definition of "Safety" in the Oxford Dictionary is the state of being safe and protected from danger or harm.

The definition of "Safety" in International Civil Aviation Organization (ICAO) Annex 19 is the state in which risks associated with aviation activities, and related to, or in direct support of the operation of aircraft, are reduced to and controlled at an acceptable level.

Safety risk: The predicted probability and severity of the consequences or outcomes of a hazard.

1.2.3 Failure Condition Classifications and Probability Terms

Failure Condition: A condition having an effect on the airplane and/or its occupants, either a direct or a consequential, which is caused or contributed to by one or more failures or errors, consideration flight phase and relevant adverse operational or environmental conditions or external events [2].

1. Classifications

   Failure Conditions may be classified according to the severity of their effects as follows [2]:

   a. *No Safety Effect*: Failure Conditions that would have no effect on safety; e.g., Failure Conditions that would not affect the operational capability of the airplane or increase the crew workload.

   b. *Minor*: Failure Conditions that would not significantly reduce the airplane safety and that involve crew actions that are well within their capabilities. Minor Failure Conditions may include, e.g., a slight reduction in safety margins or functional capabilities and a slight increase in crew workload, such as routine flight plan changes, or some physical discomfort to passengers or the cabin crew.

   c. *Major*: Failure Conditions that would reduce the capability of the airplane or the ability of the crew to cope with adverse operating conditions to the extent that there would be, e.g., a significant reduction in safety margins or functional capabilities, a significant increase in crew workload or decrease of crew efficiency, discomfort to the flight crew, or physical distress to passengers or cabin crew, possibly including injuries.

   d. *Hazardous*: Failure Conditions that would reduce the capability of the airplane or the ability of the crew to cope with adverse operations and conditions to the extent that there would be:

      i. a large reduction in safety margins or functional capabilities;

      ii. physical distress or excessive workload such that the flight crew cannot perform their tasks accurately or completely; or

      iii. serious or fatal injury to a relatively small number of occupants other than the flight crew.

   e. *Catastrophic*: Failure Conditions that would result in multiple fatalities, usually with the loss of the airplane. (Note: A "Catastrophic" Failure Condition was defined in previous versions of the rules and the advisory materials as a Failure Condition that would prevent continued safe flight and landing.)

2. Qualitative Probability Terms

When using qualitative analyses to determine the compliance with §25.1309(b), the following descriptions of the probability terms used in §25.1309 have become commonly accepted as aids to engineering judgment [2]:

   a. Probable Failure Conditions are those anticipated to occur one or more times during the entire operational life of each airplane.

   b. Remote Failure Conditions are those that are unlikely to occur to each airplane during its total life but may occur several times when considering the total operational life of a number of airplanes of the same type.

   c. Extremely Remote Failure Conditions are those not anticipated to occur to each airplane during its total life but which may occur a few times when considering the total operational life of all airplanes of the type.

    d. Extremely Improbable Failure Conditions are those so unlikely that they are not anticipated to occur during the entire operational life of all airplanes of one type.

3. Quantitative Probability Terms

When using quantitative analyses to help determine compliance with 25.1309(b), the following descriptions of the probability terms have become commonly accepted as aids to engineering judgment. They are expressed in terms of acceptable ranges for the Average Probability Per Flight Hour [2].

    a. Probability Ranges

        i. Probable Failure Conditions are those having an Average Probability Per Flight Hour greater than the order of 1×10^{-5}.

        ii. Remote Failure Conditions are those having an equal Average Probability Per Flight Hour of the order of 1×10^{-5} or less but greater than the order of 1×10^{-7}.

        iii. Extremely Remote Failure Conditions are those having an Average Probability Per Flight Hour of the order of 1×10^{-7} or less but greater than the order of 1×10^{-9}.

        iv. Extremely Improbable Failure Conditions are those having an Average Probability Per Flight Hour of the order of 1×10^{-9} or less.

1.3 EVOLUTION HISTORY OF THE AIRCRAFT SAFETY ASSESSMENT PROCESS

The method of aircraft system safety assessment is closely related to that of aircraft safety design, and thus analyzing the evolution of the aircraft design means is helpful to learn the evolution history of the aircraft safety assessment process.

The means of aircraft safety design has undergone evolutionary processes from absolute safety to the fail-safe design concept, which can be roughly divided into four periods.

1. Integrity Integrity was the first design concept with regard to aircraft system safety from 1900 to 1930. The design principle is try the best to make good components and integral system functions. This concept was an original and formed at the beginning of the aviation industry, and it has been a persistent pursuit of aviation pioneers, at the cost of their lives for some. Despite many efforts having been made to improve the integrity of aircraft, many failures have occurred along with the gradual increase of exposure duration of aircraft operation. In particular, with the occurrence of many unexpected single failures, the uncertainty of aircraft flight safety has increased. In this period, the aircraft could not be widely used for commercial operations. The representative aircrafts during this period were: the Wright Flyer in 1903, St. Louis Wraith Aircraft in 1927, and Ford three-engine Aircraft in 1930.

2. **Integrity with Selective Design Features** During 1930 to 1945, integrity was not enough to ensure flight safety. Thus, integrity with selective redundancy for some limited design features was used to further improve the safety. The concept was put into practice during World War II. In this period, the redundancy design was applied to the engine, and systems such as the radios and airspeed indicators of aircraft, and the failure rate of single failure was calculated as well. Because of the low level of safety, the aircraft in this period could not gain the public's trust; problems existed in the flight control system, propeller, and engine fire. The representative aircrafts during this period were mainly transport category airplanes serving in World War II, such as the Douglas DC-3, DC-4, and Beechcraft-18.

3. **Single Failure Concept** From 1945 to 1950, the demand for military aircraft declined sharply, which turned the attention of the aviation industry to civil aviation. To facilitate the development of civil aviation, the safety of aviation flights needed to be further improved. Therefore, the industrial community and American government formulated the "single failure concept" in 1945. This concept concentrates on protection from catastrophic single failures, i.e., at least one failure during each flight must be assumed to occur regardless of probability. This concept has a significant effect on reducing single-failure accidents. The aviation safety was improved significantly during the five years from 1945 to 1950. The extent of the public trust with regard to the flight safety increased correspondingly, resulting in the great growth in air travel. The representative aircrafts in this period were the Lockheed "Constellation" and Douglas DC-6.

Although "single failure concept" has been implemented, accidents still happen. The results of a large number of accident investigations suggested that the accidents are as a result of combinations of more than one failure.

4. **Fail–Safe Design Concept** The fail–safe design concept is proposed in 1955 and still in use today. Considering the severe losses caused by combined failures, the fail–safe concept has replaced the single failure concept for the purpose of making the aviation industry pay more attention to multifailure accidents. In 1955, the following concept was introduced for turbine-powered airplanes in the certification regulations for new transport category airplane: any single failure plus any foreseeable combination of failures must be considered. Thus, the basic principle of fail–safe design came to be that, during any flight, the combination of a single failure and foreseeable failures would not prevent aircraft from continued safe flight and landing. Although the principle of this concept remains unchanged, its implementation methods are continually improved and embodied in the application of all the three generation commercial jet airplanes.

For the earliest commercial jet aircraft applying the principle of fail–safe design, such as the Boeing707, Comet 4, DC-8, Caravelle, and Convairs, the safety test is adopted for safety verification. The safety test was derived from the design principle of "integrity". The content of the safety assessment was mainly given by means of requirements and was assessed and verified through the safety test. The safety test markedly decreases the accidents rate and further increases the trust of the public, as well as promotes the development of civil aviation.

For representative types such as Boeing727, Boeing737-00/200, Boeing747, DC-9, L-1011, DC-10, and Airbus A300, the Failure Modes and Effects Analysis (FMEA) method is formally used to perform safety design and verification when applying the principle of Fail-safe design. At the end of the 1960s, although the safety assessment was performed by safety tests during the aircraft development process, the accident rate was much higher than expected. In the 1950s, FMEA was first used for the primary flight control system by Grumman Aircraft for frequent accidents caused by oil pressure devices of fighters, which had developed into a mature phase by the mid-1960s. The FMEA method is applied for the design principle of single failure, considering both single and hidden failures. The effects of each single failure must be analyzed because it is difficult to meet the safety objectives by conducting safety tests only.

After applying FMEA, although favorable results have been achieved in lowering accident rates, accidents still occur in relevant hardware, such as combined failures in automatic flight systems.

For representative types such as all series of Boeing737 to Boeing777, MD-80, MD-90, MD-11, Airbus A319, and Airbus A340, these safety analysis methods are applied when applying the principle of Fail-safe design, such as Functional Hazard Assessment (FHA), FMEA, and Fault Tree Analysis (FTA). By using these safety analysis methods, the accident rates of relevant systems have been lowered substantially. Instead of airborne system failure, flight crew errors, maintenance errors, and other issues have become the main causes for accidents in these aircraft types.

By the 21st century, the application of complex electronic hardware and software has brought challenges to the original concept. The safety objectives could be achieved by means of "test-improvement-test" during the aircraft development process. At this point, it is necessary to identify, analyze, and control hazards throughout the entire life cycle, emphasizing that the safety requirements should be incorporated into the systems during the system development process to ensure the system safety in the processes of testing, manufacturing, operation, and maintenance. Against this background, FAA and EASA have recommended to use certain safety processes and analysis methods, including FHA, Preliminary System Safety Assessment (PSSA), System Safety Assessment (SSA), FMEA, and Common Cause Analysis (CCA), to conduct the system safety assessment.

For highly-integrated or complex systems, exhaustive testing may be impossible. For these systems, compliance may be shown by Development Assurance. Development Assurance is that all those planned and systematic actions used to substantiate, to an adequate level of confidence, that errors in requirements, design, and implementation have been identified and corrected such that the system satisfies the applicable certification basis [2].

Development Assurance Level (DAL) is the foundation of carrying out development assurance activities.

DAL is a series of levels (A, B, C, D, and E) specified for failure conditions due to the failure of the system or equipment. DAL is the foundation of carrying out development assurance activities. It is used to measure reducing errors in the function and item development process. The DAL is determined during the safety assessment process. Its purposes are to select appropriate quality assurance process during the function and item development and to develop procedures and verification standards for different levels to minimize the errors or omissions in the requirements, design and implementation.

1.4 REVISION HISTORY OF §25.1309

§25.1309 stipulates a general requirement for system safety and is applicable to any airborne equipment and systems.

1.4.1 The Content of §25.1309

The content of FAR25.1309 is as follows [3]:

FAR25.1309 Equipment, systems, and installations

(a) The equipment, systems, and installations whose functioning is required by this subchapter, must be designed to ensure that they perform their intended functions under any foreseeable operating condition.

(b) The airplane systems and associated components, considered separately and in relation to other systems, must be designed so that—

   (1) The occurrence of any failure condition which would prevent the continued safe flight and landing of the airplane is extremely improbable, and

   (2) The occurrence of any other failure conditions which would reduce the capability of the airplane or the ability of the crew to cope with adverse operating conditions is improbable.

(c) Warning information must be provided to alert the crew to unsafe system operating conditions and to enable them to take appropriate corrective action. Systems, controls, and associated monitoring and warning means must be designed to minimize crew errors which could create additional hazards.

(d) Compliance with the requirements of paragraph **b** of this section must be shown by analysis, and where necessary, by appropriate ground, flight, or simulator tests. The analysis must consider—

   (1) Possible modes of failure, including malfunctions and damage from external sources.

   (2) The probability of multiple failures and undetected failures.

   (3) The resulting effects on the airplane and occupants, considering the stage of flight and operating conditions, and

   (4) The crew warning cues, corrective action required, and the capability of detecting faults.

(Continued)

(Continued)

(e) In showing compliance with paragraphs (a) and (b) of this section with regard to the electrical system and equipment design and installation, critical environmental conditions must be considered. For electrical generation, distribution, and utilization equipment required by or used in complying with this chapter, except equipment covered by Technical Standard Orders containing environmental test procedures, the ability to provide continuous, safe service under foreseeable environmental conditions may be shown by environmental tests, design analysis, or reference to previous comparable service experience on other aircraft.

(f) EWIS must be assessed in accordance with the requirements of §25.1709.

[Amdt. 25-23, 35 FR 5679, Apr. 8, 1970, as amended by Amdt. 25-38, 41 FR 55467, Dec. 20, 1976; Amdt. 25-41, 42 FR 36970, July 18, 1977; Amdt. 25-123, 72 FR 63405, Nov. 8, 2007].

The content of CS25.1309 is as follows [4].

CS 25.1309 Equipment, systems and installations
  (See AMC 25.1309)
  The requirements of this paragraph, except as identified below, are applicable, in addition to specific design requirements of CS-25, to any equipment or system as installed in the aeroplane. Although this paragraph does not apply to the performance and flight characteristic requirements of Subpart B and the structural requirements of Subparts C and D, it does apply to any system on which compliance with any of those requirements is dependent. Certain single failures or jams covered by CS 25.671(c)(1) and CS 25.671(c)(3) are excepted from the requirements of CS 25.1309(b)(1)(ii). Certain single failures covered by CS 25.735(b) are excepted from the requirements of CS 25.1309(b). The failure effects covered by CS25.810(a)(1)(v) and CS 25.812 are excepted from the requirements of CS 25.1309(b). The requirements of CS 25.1309(b) apply to powerplant installations as specified in CS 25.901(c).

(a) The aeroplane equipment and systems must be designed and installed so that
  (1) Those required for type certification or by operating rules, or whose improper functioning would reduce safety, perform as intended under the aeroplane operating and environmental conditions.
  (2) Other equipment and systems are not a source of danger in themselves and do not adversely affect the proper functioning of those covered by sub-paragraph (a)(1) of this paragraph.

(Continued)

(Continued)

(b) The aeroplane systems and associated components, considered separately and in relation to other systems, must be designed so that—

   (1) Any catastrophic failure condition

      (i) is extremely improbable; and

      (ii) does not result from a single failure; and

   (2) Any hazardous failure condition is extremely remote; and

   (3) Any major failure condition is remote.

(c) Information concerning unsafe system operating conditions must be provided to the crew to enable them to take appropriate corrective action. A warning indication must be provided if immediate corrective action is required. Systems and controls, including indications and annunciations must be designed to minimize crew errors, which could create additional hazards.

(d) Electrical wiring interconnection systems must be assessed in accordance with the requirements of CS 25.1709.

   [Amdt No: 25/5]

   [Amdt No: 25/6]

The FAA authorized the Aviation Rulemaking Advisory Committee (ARAC) to revise the FAR25.1309 in 2002. The changes proposed were developed in cooperation with the Joint Aviation Authorities (JAA) of Europe (now the EASA) and the ARAC. The ARAC completed the proposed §25.1309 and the AC 25.1309-1B (Arsenal) in June 2002. The proposed content of §25.1309 is identical to that of CS25.1309 listed above (there may be some editorial changes). However, FAA has not yet adopted the proposed §25.1309 and AC 25.1309-1B (Arsenal) until now.

The differences between FAR25.1309 and CS25.1309 are as follows:

The description of FAR25.1309 (a) is consistent with CS25.1309 (a) (1), but CS25.1309 (a) adds the sub-paragraph (a) (2), and therefore, the requirements of the equipment and systems of those not covered by sub-paragraph (a)(1).

The failure condition probability requirements are listed in FAR25.1309 (b) when the occurrence of preventing "the continued safe flight and landing of the airplane" or reducing "the capability of the airplane or the ability of the crew to cope with adverse operating conditions." However, CS25.1309 paragraph (b) provides the requirements of occurrence probability when any failure condition is catastrophic, hazardous, or major with the emphasis that any catastrophic failure condition does not result from a single failure.

There is no difference between FAR25.1309 (c) and CS25.1309 (c).

FAR25.1309 (d) describes four aspects that need to be considered when showing compliance with the requirements of paragraph (b). The content is not included in CS25.1309.

FAR25.1309 (e) emphasizes that the critical environmental conditions must be considered, while this content is not included in CS25.1309.

FAR25.1309 (f) corresponds to CS25.1309 paragraph (d), and the descriptions of both are consistent.

1.4.2 The Origin of Safety Objectives

The aviation industry recognized as early as the late 1950s that rational acceptable quantitative probability values would have to be established. In the 1960s, quantitative probability gained in popularity, and acceptance was taken as a tool for objectifying engineering judgments.

The primary objective in establishing these guidelines was to ensure that the proliferation of critical systems would not increase the probability of serious accidents. Historical evidence at the time indicated that the probability of serious accidents due to operational and airframe-related causes was approximately one (accident) per million hours of flight. Further, approximately 10% of the total accidents were attributed to failure conditions caused by systems of airplane. Consequently, it was determined that the probability of serious accidents resulting from all such failure conditions should not be greater than one per 10 million flight hours, or "1×10^{-7} per flight hour," for a newly designed airplane. Commensurately greater acceptable probabilities were established for less severe outcomes.

The difficulty with the 1×10^{-7} per flight hour probability of a serious accident, was that all the systems on the airplane must be analyzed numerically before it was possible to determine whether the target had been met. For this reason, the (somewhat arbitrary) assumption can be made that there are about 100 failure conditions in a transport category airplane which could be catastrophic. It apparently was also assumed that, by regulating the frequency of less severe outcomes:

- only "catastrophic failure conditions" would significantly contribute to the probability of catastrophe, and
- all contributing failure conditions could be foreseen.

Therefore, the targeted allowable average probability per flight hour of 1×10^{-7} was thus apportioned equally among 100 catastrophic failure conditions, resulting in an allocation of not greater than 1×10^{-9} to each. The upper limit for the average probability per flight hour for catastrophic failure conditions became the familiar "1×10^{-9}." Failure conditions having less severe effects could be relatively more likely to occur.

FAA adopted these guidelines in AC25.1309-1, "System Design Analysis" (dated Sep. 7, 1982). This AC established an approximate probability value for the term "extremely improbable" as used in §25.1309(b), as well as the other relevant probability terms.

The intent of the term "Extremely Improbable": the objective of using this term in the regulations has been to describe a condition (usually a failure condition) that has a probability of occurrence so remote that it is not anticipated to occur in service on any transport category airplane to which the standard applies. However, while a rule sets a minimum standard for all the airplanes to which it applies, compliance determinations are limited to individual type designs. Consequently, in practice, all that has been required of applicants is a sufficiently conservative demonstration that a condition is not anticipated to occur in service during the entire operational life of all airplanes of the type design being assessed. Experience indicates that the level of conservatism traditionally provided in proper safety assessments more than compensates for the cumulative risk effects across airplane types.

The means of demonstrating that the occurrence of an event is "extremely improbable" varies widely depending on the type of system, component, or situation that must be assessed. Failure conditions resulting from a single failure are not considered "extremely improbable," and thus probability assessments normally involve failure conditions resulting from combined failures. Both qualitative and quantitative assessments are used in practice, and both are often necessary to some degree to support a conclusion that an event is "extremely improbable." Generally, performing only a quantitative analysis to demonstrate that a failure condition is extremely improbable is insufficient due to the variability and uncertainty in the analytical process. Any analysis used as evidence that a failure condition is extremely improbable should include justification of any assumptions made, data sources and analytical techniques to account for the variability and uncertainty in the analytical process. Refer to AC25.1309-1B (Arsenal), or a later revision, for acceptable means of compliance. In short, wherever part 25 requires that a failure condition be "extremely improbable," the verification means (whether qualitative, quantitative, or a combination of the two), along with engineering judgment, must provide convincing evidence that the failure condition should not occur in service [2].

1.4.3 Revision History of §25.1309

This book takes FAR25.1309 as an example, presenting the revision process of the regulation. In 1965, Regulation of Civil Aviation of Federal CAR4.606 changed into FAR25.949, and then the FAA changed FAR25.949 into FAR25.1309 during the following process of uniformly changing the rule numbering plan, putting forward a general requirement in safety.

§FAR25.1309 has been modified five times since 1965; the revision history is shown in Table 1.1. FAR25.1309 is under a new revision now.

Amendment 25-0: CAR4b.606 was changed into FAR25.949 in 1965. Then, FAR25.949 was changed into FAR25.1309 by the FAA in the progress of uniformly changing the rule numbering plan and putting forward the overall safety requirements [5].

Amendment 25-23: As experiences show, in a complex system consisting of many components, more than one failure may occur in one flight. Therefore, combined failures shall be taken into consideration to ensure adequate reliability, redundancy, and separation. Experiences also show that it is necessary to give consideration to the warnings, system controls, and operational procedures, which can minimize crew errors;

Table 1.1 The revision history of FAR25.1309

REF	Amendment	Final rule	Effective date	NPRM
1	25-0	Recodification and new Part 25	1965.02.01	64-28
2	25-23	Transport Category Airplane Type Certification Standards	1970.05.08	68-18
3	25-38	Airworthiness Review Program, Amendment No. 3: Miscellaneous Amendments	1977.02.01	75-10
4	25-41	Airworthiness Review Program, Amendment No. 5; Equipment and Systems Amendments	1977.09.01	75-10 75-23
5	25-123	Enhanced Airworthiness Program for Airplane Systems/Fuel Tank Safety (EAPAS/FTS)	2007.12.10	05-08

meanwhile, it is suggested to adopt appropriate measures to support a comprehensive failure analysis of system to meet the safety objectives. This amendment has strengthened the fail-safe design principle by modifying §25.1309(c), requiring that the warnings be given to crews in time for proper recoveries to be adopted to solve the unsafe working condition of the system. By modifying §25.1309(b) and (d), the amendment clarifies that it is impossible to remove all hazardous systems, and therefore the severity of the failure condition and the allowable probability should be addressed to control the risks [6].

Amendment 25-38: This revision mainly targeted the verbal expression of regulations, and there is no substantive change in terms of content [7].

Amendment 25-41: This revision focused on the scope and rigor of the content and changed the content "…must be designed to minimize the possibility of crew errors that would create additional hazards" of paragraph (c) to the content "… must be designed to minimize crew errors which could create additional hazards." The revision of paragraph (c) emphasizes the required design considerations of minimizing crew errors more appropriately [8].

Amendment 25-123: This revision was mainly to tie in with the H subpart, which emphasizes the EWIS safety requirements. Meanwhile, the original paragraph (e) was moved to the newly added §25.1310 [9].

1.5 COMPLIANCE WITH §25.1309

The §25.1309 is a regulation, while the AC is suggested and explanatory materials for describing acceptable compliance means, but not the only means for demonstrating compliance with the applicable regulations. AC25.1309-1A/1B (Arsenal) provides related definitions (e.g., the classification of failure conditions, etc.), the safety objectives and the specific means of compliance for §25.1309. The suggested means to demonstrate compliance with §25.1309 given by the certification authorities in AC25.1309-1A/1B (Arsenal) is neither mandatory nor regulatory.

Some industry standards have been referred to or accepted in the AC, while most of them have not. In the actual Type Certification (TC) process, it is necessary to use the TC documents (e.g., the issue papers or compliance documents) for affirmation whether the industry standards adopted have been referred to by the AC. If the related industry standard has been referred to by the AC, the applicant is usually advised to adopt that standard.

AC25.1309-1B (Arsenal) refers to many industry standards, namely, SAE ARP4754, SAE ARP4761, Radio Technical Commission for Aeronautics (RTCA) DO-178B and DO-160.

Note: the AC 25.1309-1B (Arsenal) released in 2002, and the RTCA DO-254 was adopted by FAA in 2005.

In regard to §25.1309, the relationship among regulations, AC, and the industry standards is shown in Fig. 1.1.

The industry standards referred to in Fig. 1.1 will be described in subsequent sections. It is necessary to introduce the acceptance of the FAA to these industry standards.

The first industry standard in Fig. 1.1 accepted by the FAA was RTCA DO-178. The FAA accepted RTCA DO-178B by issuing AC20-115B in Jan. 1993. The European airworthiness authority (called JAA at that time) accepted Temporary Guidance Leaflet 4 as the compliance means for airborne software. Other airworthiness authorities also used similar methods to accept the standard, e.g., CAAC approved RTCA DO-178B as the compliance means for system and equipment software to meet the airworthiness requirements through CAAC AC21-02 Software

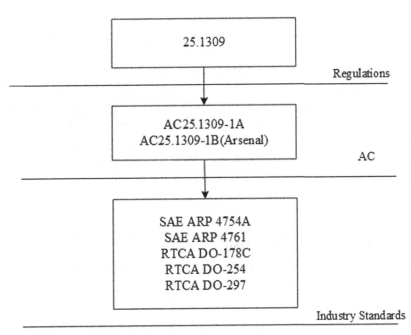

Figure 1.1 The relationship between regulations, AC, and industry standards of §25.1309.

Review Method for airborne System and equipment Certification. Moreover, FAA also accepted DO–178B as the compliance means for §25.1309 in AC25.1309–1B (Arsenal).

RTCA/DO–178C series standards were officially approved as the compliance means for airborne software in Jul. 2013 through AC 20–115C by FAA. The series include DO–178C Software Considerations in the Certification of Airborne Systems and Equipment, DO–330 Considerations on Software Tools Identification, DO–331 Supplemental Document for Model-based Development and Verification of DO–178C and DO–278A, DO–332 Supplemental Document for Target Technology and Related Technology-Oriented of DO–178C and DO–278A, DO–333 Supplementary Document for Formal Method of D0–178C and DO–278A, etc.

AC25.1309–1B (Arsenal) refers to SAE ARP 4754 Certification Consideration on Highly comprehensive and Complex System (release in Nov., 1996). Though AC25.1309–1B (Arsenal) has not been officially released, FAA allows its certificate personnel to use the main content in the AC in the form of a letter, including the approval of reference to SAE ARP 4754.

However, after the publication of SAE ARP 4754A, FAA issued AC20–174 (the first version without previous revision) for approval of this standard.

DO–254 was released in 2000. FAA released AC20–152 in 2005, which indicated its official approval of DO–254; the DO–254 approval can be regarded as the compliance means for a complex programmable electronic device with its development assurance level of A, B, or C, including Application Specific Integrated Circuit (ASIC), Programmable Logic Device (PLD), Field Programmable Gate Array (FPGA), and other similar electronic devices. Other airworthiness authorities also approve the standard by using similar methods.

1.5.1 AC25.1309-1B (Arsenal)

AC25.1309–1B (Arsenal) describes acceptable means for showing compliance with §25.1309. The effective edition is AC25.1309–1A, which was issued on Jun. 21, 1988. AC25.1309–1B (Arsenal) has not been effective yet. However, AC 25.1309–1B (Arsenal) has been used by most of the applicants and authorities of the world for many years. The revision of AC25.1309–1 is shown in Table 1.2.

Table 1.2 The revision of AC 25.1309-1

Ref	AC no	Subject	Effective date
1	AC25.1309-1	System Design Analysis	Sep. 7, 1982
2	AC25.1309-1A	System Design and Analysis	Jun. 21, 1988
3	AC25.1309-1B (Arsenal)	System Design and Analysis	Drafted in 2002

AC25.1309-1B (Arsenal) defines the quantitative average probability value, such as 1×10^{-9} per flight hour. The term "Average Risk" is to represent the average probability of failure for some baseline airplanes over the total life. This average risk sets a threshold requirement for the overall fleet but not for individual airplanes or flights. Conceptually, it is quite possible for an aircraft fleet to meet average risk criteria even if individual airplanes (or individual flights) are at undesirable risk levels.

FAA is evolving the industry by convening the ASAWG (Airplane-level Safety Analysis Working Group) to address specific risks. The definition of a specific risk is: "The risk on a given flight due to a particular condition." Accordingly, the definition of a Specific Risk of Concern is: when the airplane is one failure away from a catastrophe, or when the risk is greater than the average probability criteria provided in AC25.1309-1B (Arsenal) for hazardous and catastrophic failure conditions, on a given flight due to a particular condition. The particular conditions identified for detailed consideration are latent failure, Master Minimum Equipment List (MMEL), the flight and diversion time, etc. The intended outcome is the release of AC25.1309-1B (Arsenal).

However, FAA genaraly adopts AC25.1309-1B (Arsenal) as acceptable compliance means, except §8(d). The NOTICE in Federal Register (FR) Volume 68 No. 82 *Advice on the Change of the Standards and Advice of the System Design and Analysis*, released on Apr. 29, 2003, showed that §8(d) in AC25.1309-1B (Arsenal) should not be regarded as the acceptable compliance means because §8(d) describes an alternative method that should be dealt with in accordance with the exemption procedure and is intended to be removed from AC25.1309-1B [10]. The reason is probably that the §8(d) lowers the requirement of "the average probability per flight hour of each catastrophic failure conditions must be equal to or less than 1×10^{-9}."

1.5.2 Industry Standards Related to §25.1309

1.5.2.1 SAE ARP 4754 and 4754A

SAE ARP 4754, *Certification Considerations for Highly Integrated or Complex Aircraft Systems*, was issued in 1996. It is a successful application of system engineering in civil aircraft, and it has been used by the aviation industry for almost 20 years. It concludes the best engineering practices in the aviation industry from the top-down system development process. It also offers the tasks and elements that relate to airworthiness certification during the integral process, which relates to safety assessment, requirements capture, validation, verification, configuration management, process assurance, and certification coordination. In this way, it will provide the compliance demonstration support for airworthiness certification.

ARP4754 complies with the system engineering manuals issued by the International Association of Systems Engineers. However, it mainly focuses on the technical activities related to airworthiness safety in the development of the system, with less involvement in project management, project-enabling processes, and agreement processes [11].

Out of consideration of the new challenges that come from the development of new technologies, SAE started the revision of ARP4754 in 2003 and issued SAE ARP 4754A in 2010. Based on ARP4754, 4754A made the following changes [12].

1. Adding a description of how to meet the certification-related tasks and elements to form a compliance means is more instructive.
2. In view of the successful application of ARP4754, the scope of application will be expanded from "highly integrated and complex system" to "civil aircraft and system."
3. In recent years, for the development of new technologies and standards, it absorbed RTCA DO-297 and SAE ARP 5150/5151, which cover the development and operation of the entire lifecycle of safety activities and are suitable for Integrated Modular Avionics (IMA) applications in civil aircraft.
4. Highlighting the importance of developing the plans and optimizing the DAL assignment logic to better guide applicants and improve aircraft safety.

In September 2011, the FAA issued AC 20-174 to formally approve SAE ARP 4754A. Therefore, it is an important way to get recognition from the airworthiness authority by developing the aircraft or airborne systems in full accordance with ARP4754A.

In ARP4754A, the safety process in SAE ARP 4761 is regarded as part of the development process, and it has a close interaction with other development activities and integral activities. In addition, during the safety assessment, we also need to complete the safety requirements generation, safety requirements validation, safety requirements verification, configuration management, process assurance, and certification coordination.

This book will mainly focus on the system safety assessment, and it will involve the validation and verification of safety requirements. The configuration management, process assurance, and certification coordination will not be described in this book.

1.5.2.2 SAE ARP 4761

SAE ARP 4761, *Guidelines and Methods for Conducting the Safety Assessment Process on Civil Airborne Systems and Equipment,* issued in December 1996, describes guidelines and methods of performing the safety assessment for civil aircraft. It is primarily associated with showing compliance with §25.1309.

This document presents guidelines for conducting an industry-accepted safety assessment consisting of FHA, PSSA, and SSA. It also presents information on the safety analysis methods needed to conduct the safety assessment. These methods include FTA, Dependence Diagram (DD), Markov Analysis (MA), FMEA, FMES, ZSA, PRA, and CMA. Appendix L provides a contiguous safety assessment process example of Wheel Brake System. This contiguous example illustrates safety assessment process and methods of an aircraft or system through the overall development cycle.

The SAE S-18 committee has revised the SAE ARP 4761 for several years. There may be some changes in the new edition, SAE ARP 4761A.

- Newly added two safety assessment processes: Preliminary Aircraft Safety Assessment (PASA) and Aircraft Safety Assessment (ASA).
- Newly added model-based safety analysis.
- Newly added cascading failure analysis.
- Move the DAL assignment from 4754 A to 4761 A.

1.5.2.3 RTCA DO-178C

In the early 1970s, with the large-scale use of software in the airborne system and equipment, the airborne software has a great influence on the safety of the equipment, system, and aircraft. To resolve the problem, it is urgent to build a set of standards to ensure that the airborne software meets the airworthiness requirement of the safety level.

In May 1980, RTCA, Inc. established the "Digital Avionics Software" Committee, which is dedicated to developing and documenting software practices to support the development of software-based airborne systems and equipment. In January 1982, RTCA and EUROCAE completed the work and formally published DO-178, "Software Considerations in Airborne Systems and Equipment Certification."

In 1983, RTCA Executive Committee determined that DO-178 should be revised to reflect the experience gained in the certification of the aircraft and engines containing software-based systems and equipment. Thus, RTCA and EUROCAE completed the work and published DO-178A, "Software Considerations in Airborne Systems and Equipment Certification," in 1985.

With the rapid development of software technology and different understanding of some critical issues, it is necessary to revise DO-178A. In 1989, the FAA formally requested that RTCA establish a special committee to revise DO-178A. Thus, in December 1992, RTCA and EUROCAE formally completed DO-178B, "Software Considerations in Airborne Systems and Equipment Certification."

In 2004, FAA and representatives of the aviation industry initiated a discussion with RTCA concerning the advances in software technology since 1992, when DO-178B/ED-12B was published. RTCA then requested a Software Ad Hoc committee to evaluate the issues and determine the necessity for improved guidance in light of these advancements in technology. Thus, RTCA and EUROCAE carried out the revision activities and published DO-178C series documents in December 2011.

AC 20-115C issued by FAA accepted DO-178C series documents as an acceptable compliance means of airborne software in July 2013. DO-178C series documents consist of DO-178C, *Software Considerations in Airborne Systems and Equipment Certification*; DO-330, *Software Tool Qualification Considerations*; DO-331, *Model-Based Development and Verification Supplement to DO-178C and DO-278A*; DO-332, *Object-Oriented Technology and Related Techniques Supplement to DO-178C and DO-278A*; and DO-333, *Formal Methods Supplement to DO-178C and DO-278A* [13].

Now FAA, EASA, and other certification authorities accept DO-178C series documents as airborne software design and verification standard in civil aviation. The documents are used to guide the software development process of airborne equipment and systems and ensure that the software completes its intended function and conforms to the airworthiness requirements. The civil aviation industry uses the documents as the compliance means of airborne software. The civil aviation industry develops software life cycle processes in accordance with documents, satisfies objectives for software life cycle processes, accomplishes the activities that provide a means for satisfying those objectives, and provides software life cycle data that indicates that the objectives have been satisfied.

1.5.2.4 RTCA DO-254

RTCA/DO-254, *Design Assurance Guidance for Airborne Electronic Hardware*, is a document for design and verification of complex airborne electronic hardware that is ratified by FAA, EASA, CAAC, and other certification authorities around the world. The main purpose of this standard is to ensure the safety of airborne electronic hardware equipment through the implementation of this standard.

DO-254 was issued in 2000 by RTCA and EUROCAE and was officially ratified by the FAA in 2005. The document provides guidance for design and assurance of complex electronic hardware applied in airborne systems and equipment. Thus, the companies, which design electronic products of airborne equipment, must follow the DO-254 during their FPGA and ASIC design process. At the same time, in the military, space and defense, and other areas, the standard is also recommended. Other related industries, such as musical instruments and nuclear power, are also looking for a standard similar to DO-254.

A few years ago, airborne electronic hardware was considered to be "simple" or "can be verified by system level test." Therefore, the hardware verification is relatively easy. This result has led to many functions being implemented through transferring from software to hardware. At the same time, the hardware design becomes more and more complicated, e.g., in PLD and ASIC, the embedded logic is more complex than the software. Because FAA, EASA, and certification authorities desire to ensure that complex airborne electronic hardware can work reliably as expected, they require hardware equipment suppliers to reach DO-254 to avoid incorrect operation and potential aviation disasters.

DO-254 has put forward a series of objectives for the design and supporting processess of hardware. The design process includes requirements capture, conceptual design, detailed design, implementation and production transition process, etc. The support process includes validation, verification, configuration management, and process assurance and certification liaison process. DO-254 is used as a compliance mean in the manufacturing process of the Boeing B787 and Airbus A380 for airborne electronic hardware. Through industry practices, the adoption of DO-254 in actual projects can yield many benefits, for example [14]:

- making design requirements clear;
- improving the assurance process of products;
- finding design flaws as soon as possible;
- ensuring consistency between hardware implementations and requirements;
- improving configuration management.

However, it is found that there are still some shortcomings of DO-254 in practical application, for example:

- the given work instructions are written at equipment layer, but its usage is limited to PLD level;
- the requirements of verification are too simple, and the advanced verification method is difficult to understand;
- the guidance of COTS is not comprehensive enough;
- the standard is a little bit out of line with the current electronic hardware design and verification technology.

1.5.2.5 RTCA DO-297

RTCA DO-297, *Integrated Modular Avionics Development Guidance and Certification Considerations*, was jointly prepared by the RTCA Special Committee (SC-200) and EUROCAE Working Group 60 and approved by the RTCA program management committee on Nov. 8, 2005.

The usage of Integrated Modular Avionics (IMA) is rapidly expanding and is found in many different types of aircraft. Because of the rapid growth, RTCA established Special Committee 200 (SC-200) and EUROCAE established Working Group 60 (WG-60) to jointly develop a document that could be used as a guidance in the design, development, and application of IMA.

This document contains guidance for IMA developers, application developers, integrators, certification applicants, and those involved in the approval and continued airworthiness of IMA systems. The guidance describes the objectives, processes, and activities for those involved in the development and integration of IMA modules, applications, and systems to incrementally accumulate design assurance toward the installation and approval of an IMA system on an approved aviation product as differentiated from traditional federated aviation system architectures.

Six tasks defying the incremental acceptance of IMA systems in the certification process are as follows [15]:

- Task 1: Module acceptance.
- Task 2: Application software or hardware acceptance.
- Task 3: IMA system acceptance.
- Task 4: Aircraft integration of IMA system—including Validation and Verification (V&V).
- Task 5: Change of modules or applications.
- Task 6: Reuse of modules or applications.

This document describes application properties as they relate to their integration with a platform; however, it does not address the specific functionality of applications or specific TSO/ETSO requirements, Minimum Operational Performance Specifications (MOPS).

1.5.2.6 RTCA DO-160G

This document defines a series of minimum standard environmental test conditions (categories) and applicable test procedures for airborne equipment. The purpose of these tests is to provide a laboratory means of determining the performance characteristics of airborne equipment in environmental conditions representative of those that may be encountered in airborne operation of the equipment.

The standard environmental test conditions and test procedures may be used in conjunction with applicable equipment performance standards as a minimum specification under environmental conditions, which can ensure a sufficient degree of confidence in performance during operations.

DO-160 (or its precursor, DO-138) has been used as a standard for environmental qualification test since 1958. It has been referenced in MOPS for specific equipment designs and is referenced in AC as a means of environmental qualification for TSO authorization.

DO-160 will continue to be subjected to revisions as needs arise in the aviation community. The latest version is DO-160G (Dec. 2010). The previous versions include DO-160A (Jan. 1980), DO-160B (Jul. 1984), DO-160C (Dec. 1989), DO-160D (Jul. 1997), DO-160 D Change 1 (Dec. 2000), DO-160 D Change 2 (Jul. 2001), DO-160 D Change 3 (Dec. 2002), DO-160E (Dec. 2004), and DO-160F (Dec. 2007).

There are 23 environmental test items contained in DO–160G [16]: Temperature and altitude, temperature variation, humidity, operational shocks and crash safety, vibration, explosive atmosphere, waterproofness, fluids susceptibility, sand and dust, fungus resistance, salt fog, magnetic effect, power input, voltage spike, audio frequency conducted susceptibility—power inputs, induced signal susceptibility, radio frequency susceptibility (radiated and conducted), emission of radio frequency energy, lightning-induced transient susceptibility, lightning direct effects, icing, electrostatic discharge, and fire, flammability.

REFERENCES

[1] SAE ARP 4761. Guidelines and methods for conducting the safety assessment process on civil airborne systems and equipment. SAE International; 1996.
[2] FAA AC 25.1309-1B (Arsenal). System design and analysis. FAA; 2002.
[3] FAR25. Airworthiness standards: transport category airplane. FAA.
[4] CS-25/Amendment 18, Certification Specifications and Acceptable Means of Compliance for Large Aeroplanes. EASA; June 22, 2016.
[5] Final Rule, Recodification and new Part 25, Docket No. 5066; Amendment No. 25-0. Federal Register; December 24, 1964. p. 18289.
[6] Final Rule, Transport Category Airplane Type Certification Standards, Docket No. 9079; Amendment No. 25-23. Federal Register; April 8, 1970 (Volume 35, Number 68). p. 5665.
[7] Final Rule, Airworthiness Review Program, Amendment No. 3: Miscellaneous Amendments, Docket No. 14324; Amendment Nos. 21-44, 23-17, 25-38, 27-11, 29-12, 31-3, 33-7, and 35-3; Federal Register; December 20, 1976 (Volume 41, Number 245). p. 55454.
[8] Final Rule, Airworthiness Review Program, Amendment No. 5; Equipment and Systems Amendments, Docket No. 14625; Amendment Nos. 23-20, 25-41, 27-13, 29-14. Federal Register; July 18, 1977 (Volume 42, Number 137). p. 36960.
[9] Final Rule, Enhanced Airworthiness Program for Airplane Systems/Fuel Tank Safety (EAPAS/FTS), Docket No. FAA-2004-18379; Amendment Nos. 1-60, 21-90, 25-123, 26-0,91-297, 121-336, 125-53, 129-43. Federal Register; November 8, 2007 (Volume 72, Number 216). p. 63363–414.
[10] Notice, Advice on the change of the standards and advice of the system design and analysis. Federal Register; 2003;68(82).
[11] SAE ARP 4754. Certification considerations for highly integrated or complex aircraft systems. SAE International; 1996.
[12] SAE ARP 4754A. Guidelines for development of civil aircraft and systems. SAE International; 2010.
[13] RTCA DO-178C. Software considerations in airborne systems and equipment certification. RTCA; 2011.
[14] RTCA/DO-254. Design assurance guidance for airborne electronic hardware. RTCA; 2005.
[15] RTCA DO-297. Integrated modular avionics development guidance and certification considerations. RTCA; 2005.
[16] RTCA DO-160G. Environmental conditions and test procedures for airborne equipment. RTCA; 2010.

CHAPTER 2

Safety Management

Contents

2.1 INTRODUCTION TO SAFETY ASSESSMENT PROCESS

Typical system development processes is iterative by using both top-down and bottom-up strategies. Generic system development process activities are as follows: aircraft function development, allocation of aircraft functions to systems, design of system architecture and allocation of requirements to items, allocation of item requirements to hardware and software, hardware and software design and integration, and system integration. Besides, the validation activities are required to ensure that the safety requirements are correct and complete, and verification activities are required to ensure that the safety objectives are met by the design.

Civil Aircraft Electrical Power System Safety Assessment
DOI: http://dx.doi.org/10.1016/B978-0-08-100721-1.00002-9
27

Safety requirements are the most important requirements during the system development process. Safety requirements should be implemented and verified by structured methods, which includes safety assessment processes of Aircraft/System Functional Hazard Assessment (AFHA/SFHA), Preliminary Aircraft/System Safety Assessment (PASA/PSSA), Aircraft/System Safety Assessment (PASA/SSA), and Common Cause Analysis (CCA). Fig. 2.1 shows the fundamental relationships between safety assessment processes and system development processes. In reality, there are many feedback loops within and among these relationships, though they have been omitted for clarity.

2.1.1 Integral Process of Safety Assessment

In the civil aircraft development, the integral process of safety assessment is a closed-loop process.

1. In the stage of conceptual design, the top-level requirements document is prepared. Meanwhile a safety plan is worked out and the common data is prepared to be used in the aircraft development process, such as the flight phases, risk time and probability of external events, and other data.

2. Based on the aircraft function list, the failure conditions and classifications are identified through the analysis on aircraft function failures. After the preliminary validation to the assumptions and Aircraft Functional Hazard Assessment (AFHA) results, the AFHA document is released. The AFHA document is one of the inputs of the aircraft functional requirements document.

3. Examine the proposed aircraft architectures and assess how their failures can lead to the aircraft-level failure conditions identified by the AFHA, and determines whether the AFHA objectives can be met. Then, the PASA is released.

4. After the allocation of aircraft functions to systems, the system safety requirements are formulated and incorporated into the system requirements document (Each system is required to have such a document.) according to the results of AFHA and that of aircraft functional requirements document/aircraft functional descriptions document.

5. The safety process continues from the aircraft level to the system level. The system functions, along with refined safety requirements, are confirmed. Each system function is analyzed in the System Functional Hazard Assessment (SFHA) based on the system functions

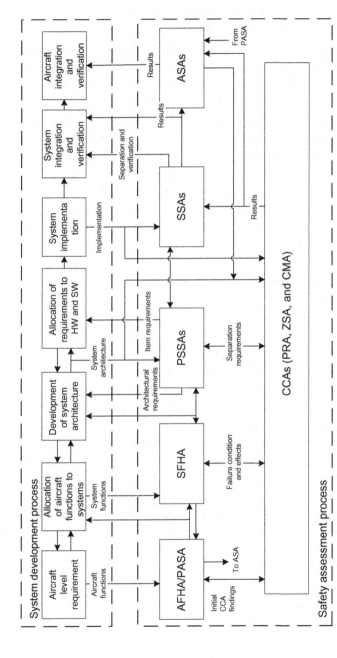

Figure 2.1 Safety assessment process

list to identify the effects and classifications of function failure conditions and safety objectives. The results of SFHA will be compared with that of AFHA and validated through the cross–check, and the results are recorded in the SFHA reports.

6. The next step is PSSA. According to the failure conditions list determined by the FHA as well as the system architecture description, qualitative and quantitative assessments are conducted for significant failure conditions, and the failure probabilities are assumed to prove that the system architecture being considered meets the safety objectives corresponding to the failure conditions. Meanwhile, the CMA at the system or item level is conducted to identify the common cause failures and/or errors that affect significant failure conditions (catastrophic and/or hazardous failure conditions), to verify the independence requirements in the design implementation. The results of these analyses are included in the PSSA report and CMA report, respectively.

   The analysis depth of the PSSA depends on the FHA results, which varies with the design, complexity, and type of the system being analyzed.

7. The safety requirements for system, installations maintenance, and equipment are the results of PSSA and CMA, which are included in the related requirements specification documents.

8. The equipment supplier conducts the safety analysis required in the Purchaser Technical Specification (PTS), such as FMEA/FMES, to demonstrate the compliance with PTS.

9. The Common Cause Analysis (CCA) is carried out through the whole safety assessment process in parallel, which includes the ZSA, PRA, and CMA. The results of CCA serve as the inputs of PSSA and SSA.

10. Based on the results of flight tests, ground tests, and simulator tests, the PSSA is updated to the firstflight PSSA. After completion of the suppliers' analyses, obtaining the CCA results and available results of flight tests and that of all validation/verification activities, the PSSA is updated to the SSA, which indicates the compliance with system level requirements. The Candidate Certification Maintenance Requirement (CCMR) is released to be used for the Certification Maintenance Requirement (CMR) process.

11. When the results of SSA for all systems are available, the aircraft level safety synthesis is conducted to demonstrate the compliance with

aircraft level requirements. In fact, the first version of aircraft level safety synthesis should be released right after the release of PSSA.

12. After the aircraft certification, the operational data are required to be analyzed for the purpose of continuous airworthiness monitoring. Based on the operational results and modifications to the aircraft, the safety assessment is updated.

13. While conducting the airworthiness monitoring, the results of products and operations should be analyzed to establish precautionary measures, and lessons learned should be collected (regulatory recommendations) for the aircraft modifications and future type design.

2.1.2 Safety Assessment Activities and Aircraft Development Life Cycle Activities

The safety assessment process supports the aircraft development life cycle activities during the whole product development process. The aim is to achieve the optimal safety level required by the operators and the airworthiness authorities. The relationship between safety assessment activities and aircraft development life cycle activities are shown in Fig. 2.2.

The safety activities that should be done at the milestones. Here, the contents of milestones are limited to safety. However, for commercial type, it should be noted that cost and market prospect are also the major factors have to be considered, which may affect the safety.

P1: Project Demonstration

The contents of P1 (limited to safety):

1. Determine project target
2. Research on market potential and prospect
3. Analysis of product development ability
4. Analysis of potential suppliers

P2: Project Start-up

1. Initiate the construction of safety assessment committee
2. Build an organizational structure of aircraft safety management
3. Initiate the work plan for all levels of safety working group include aircraft, system, suppliers, assembly, and continuous airworthiness
4. Prepare the aircraft safety plan
5. safety design methods are in place
6. Prepare safety common data document
7. Analyze the safety requirements of the markets, operators, and certification authorities

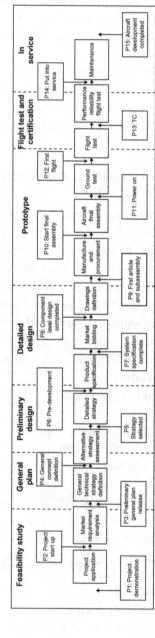

Figure 2.2 The relationship between safety assessment activities and aircraft development life cycle activities.

P2 is the beginning of safety activities of the project.

P3: Preliminary General Strategy Release

1. Initiate AFHA, identify:
   - The main functions of the aircraft and flight phases
   - Emergency and environment configuration
   - Failure conditions, effects, detection, and crew actions
   - Classification, safety objectives, supporting materials
2. Initiate PASA
   - Aircraft architecture
   - System probability budgets
   - The relationship between systems
3. Define the safety requirements or safety objectives at aircraft level
4. Initiate PRA (e.g., engine or Auxiliary Power Unit(APU) uncontained rotor failures)
5. Build safety assessment committee

P4: General Concept Definition

1. Start SFHA
   - System functions and flight phases
   - Emergency and environment configuration
   - Failure conditions, effects, detection, and crew actions
   - Classification, safety requirements/objectives, supporting materials
2. Start FTA at system level
   - System architecture
   - Subsystem probability budget
   - Intersystem relationships

P5: Strategy Selected

1. Select Strategy considering safety
   - Partitioning, dissimilarity, and monitoring
   - System architecture analysis (fault tolerance, common cause failures)
   - Analysis of failure probability
   - In-service experience evaluation
   - Installation (segregation, separation), failure independency
   - IDAL
   - Maintenance procedures and intervals
   - Failure detection and fault isolation

P6: Pre-development

1. SFHA has been agreed by the certification authorities.

2. Initiate PSSA, evaluate significant failure conditions identified in FHA and the system architecture, and allocate the safety requirements to system or subsystem, to show how the requirements be met:
   - FTA/DD/MA
   - Safety requirements allocated to lower-level system
   - Means of compliacne for qualitative and quantitative requirements
   - Supporting materials (e.g., test and analysis, etc.)
   - IDAL
   - System/hardware/software safety requirements
   - Maintenance needs
3. Initiate the preliminary CCA
   - PPA
   - ZSA
   - CMA

P7: System Specification Completed
1. Validate the safety requirements and assumptions
   - Completeness
   - Correctness
2. Define safety relevant qualification tests
3. Specify safety requirements of system/equipment
   - Qualitative and quantitative safety requirements
   - IDAL
   - Required safety analysis (e.g., FMEA/FMES, FTA)
   - Verification requirements and documentation requirements

P8: Component Level Design Completed
1. Participate in design review
2. Support the design engineering with safety activities
3. Define safety common data
4. Failure analysis and simulations
5. Electronic and electrical components failure probability allocation

P9: First Article and Subassembly
1. Start FMEA and FMES
2. Review supplier FMEA, FMES, FTA, etc.
3. Define tests derived from FMEAs
4. Participate in the verification of validated and implemented safety requirements.

P10: Start final assembly

1. ZSA
   - Divide the aircraft into several zones
   - Installation checked against independent requirements
   - Preparing report including violation and effects analysis
2. PRA
   - Risks identified
   - Simultaneous or cascading effects of each risk examined
3. CMA
   - Identify common cause failures by using checklist
   - Verify independence existance

P11: Power on

1. Initiate SSA
   - System description
   - List of failure conditions (FHA, PSSA)
   - Failure condition classification (FHA, PSSA)
   - Qualitative analyses for failure condition (FTA/DD/MA, FMES)
   - Quantitative analyses for failure condition (FTA/DD/MA, FMES)
   - CCA
   - Maintenance tasks and intervals
   - IDAL
   - Verification that safety requirements from the PSSA are realized
   - Results of the verification process

P12: First Flight

1. The safety relevant maintenance procedures and intervals have been delivered to the Maintenance Steering Group-3 (MSG-3) process
2. The safety requirements requiring test have been delivered to the test personnel
3. The PSSA have been reviewed and approved by the certification authorities.

P13: TC

1. Approval of SSAs and required CCAs, FHAs, and FMEAs by certification authorities
2. Define constraints for the MMEL and the Flight Crew Operational Manual (based on safety assessment and conclusions)
3. Approval of ASA report by certification authorities

P14: Put Into Service

Finalize the final reliability data, and provide Mean Time Between Unscheduled Removals (MTBUR) for logistics support.

P15: Aircraft Development Completed

Note: The following activities are performed after milestone P15.

1. Product improvement projects
2. Prepare safety analysis due to in-service problems
3. Update certification documents (e.g., SSA, CCA)
4. The safety relevant design changes
5. Changes of environmental and/or operational conditions
6. Changes of qualitative and quantitative assumptions
7. Prepare system safety reviews
8. Participate in reliability improvement programs
9. Monitor suppliers and verify specified and guaranteed date (e.g., MTBF, MTBUR)
10. Evaluate In-service data

2.2 SAFETY ASSESSMENT PROCESS AND ANALYSIS METHODS

2.2.1 Safety Assessment Process

The safety assessment process is a methodology to evaluate aircraft functions and the design of systems performing these functions to determine that the associated hazards for those functions have been properly addressed. The safety assessment process is qualitative and can be quantitative. According to SAE ARP4754A, the safety assessment process includes the following processes [1].

1. AFHA/SFHA: A systematic and comprehensive examination of aircraft or system functions to identify and classify the potential functional failure conditions according to the severity. The FHA is developed at the beginning of the development process and is updated as new functions or failure conditions are identified. Thus, the FHA is a living document throughout the development cycle.

2. PASA/PSSA: A systematic and comprehensive examination of a proposed aircraft or system architecture to determine how failures can lead to the aircraft or system level failure conditions. The PASA/PSSA are used to establish the requirements of lower levels and demonstrate how the requirements of the failure conditions can be met. The PASA and PSSA are updated throughout the system development process.

3. Aircraft Safety Assessment (ASA)/SSA: A systematic, comprehensive evaluation of the complete aircraft or implemented systems to show that safety objectives from the AFHA/SFHA and relevant safety requirements established by the PASA and the PSSA are satisfied.

Additionally, for appropriate management of the safety assessment process, a Safety Plan should be created.

2.2.2 Safety Analysis Methods

The safety assessment process includes safety analysis methods which may be applied throughout the development cycle to provide the analyst a means for quantitatively and qualitatively assessing the safety of a design. These methods include FTA, DD, MA, ModelBased Safety Analysis (MBSA), FMEA/ FMES, and CCA. The methods selected can vary based on system characteristics and engineering judgement. The results of these methods may stand alone or be incorporated into any of the higher level assessments.

1. FTA/DD/MA/MBSA

FTA, DD, and MA are top-down analysis methods. These analyses proceed down successively through more detailed (i.e., lower) levels of the design. After identifying the failure conditions in the FHA, the FTA/DD/MA can be applied as part of the PASA/PSSA to determine how single failure or combined failures (if any) at the lower levels might lead to the failure condition. When an FMEA/FMES is performed, a comparison should be accomplished to ensure that all significant effects identified are in the FTA/DD/MA as basic events. The failure rates of the FTA/DD/MA basic events are given by the FMEAs and/or FMES.

FTA and DD have been widely used for safety assessment as they are conceptually simple and easy to be understood. For these reasons, FTA and DD should be used whenever possible, with recognition of the following limitations [2].

1. It is difficult to allow for various types of failure modes and dependencies such as transient and intermittent failures and standby systems with spares.

2. A fault tree is constructed to assess the causes and probability of a top event. If a system has many failure conditions, separate fault trees may need to be constructed for each one of them.

In some situations it may be difficult for a fault tree to represent the system completely, such as repairable systems and systems where failure/

repair rates are state dependent. MA does not have the above indicated limitations. The sequence dependent events are included in state transition diagram, and they can cover a wide range of system behaviors.

In the classic safety assessment approach, the safety analyst manually construct fault trees or other safety analysis artifacts. The MBSA is a new methodology, and it can achieve equivalent results as the classical safety analysis methods. More details of MBSA are shown in Chapter 10, Formal Model Based Safety Analysis Methods and the Application.

2. FMEA & FMES

An FMEA is a systematic, bottom-up method of identifying the failure modes of a system, item, or function and determining the effects on the next higher level. It may be performed at any level within the system (e.g., piece-part, function, etc.). Software can also be analyzed qualitatively using a functional FMEA. Typically, an FMEA is used to address failure effects resulting from single failures.

The scope of an FMEA should be coordinated with the user requesting it. The analysis may be a piece-part FMEA or functional FMEA. If the failure rates derived from a functional FMEA allow the PSSA probability budgets to be met, a piece-part FMEA may not be necessary.

An FMES is a grouping of single failure modes that produce the same failure effects (i.e., each unique failure effect has a separate grouping of single failure modes). An FMES can be compiled from the aircraft manufacturer's, system integrator's, or equipment supplier's FMEAs. FMES is not a new analysis method, but the summary of FMEA.

3. CCA

Independence between functions, systems, or items may be required to satisfy the safety requirements. Therefore, it is necessary to ensure that such independence exists, or that the risk associated with dependence is deemed acceptable. CCA is active in parallel with any other safety assessment processes and provides methods for the evaluation on independence or the identification of specific dependencies. These parallel methods may also aid the PASA and PSSA in the generation of independence requirements (e.g., physical, installation requirements).

CCA: To establish and verify physical and functional separation, isolation, and independence requirements between systems and items, and to verify that these requirements have been met. CCA is composed of ZSA, PRA, and CMA.

ZSA: A ZSA should be performed on each zone of the aircraft. The objective of the analysis is to ensure that the equipment installation meets the safety requirements with respect to basic installation, interference between systems and maintenance errors.

PRA: Particular risks (such as fire, high energy devices, leaking fluids, and bird strike) are defined as those events or influences which are external to the aicraft or within the aircraft but external to the systems and equipment concerned, but may violate independence claims.

CMA: A CMA is performed to verify that ANDed events in the FTA/DD/MA are independent in the actual implementation. The effects of design, manufacturing, maintenance errors and failures of system components which defeat their independence should be analyzed.

2.3 MANAGEMENT OF SAFETY ACTIVITIES

During the aircraft development cycle, apart from skillfully mastering and applying the safety analysis methods, the safety activities should be managed rationally by formulating management procedures to ensure them to be implemented and carried out in a correct, timely, and effective manner.

2.3.1 Safety Program and Plan

1. Safety program

Safety program reflects the safety design culture of the company. It is a part of the design assurance system. The safety program is not applicable to any specific project. The safety program is the very top-level document and fundamental basis of the safety activities, as well as the specification guiding the aircraft safety activities to be performed scientifically and orderly.

2. Safety plan at aircraft level

In order to ensure the safety assessment activities to be implemented timely and effectively, the safety plan at aircraft level need to be compiled in accordance with the aircraft development plan, certification plan, and safety program. The project manager of the company is responsible for safety plan that is applicable to the specific project. The safety plan at the aircraft level aims at refining responsibilities and defining working breakdown, along with coordinating the aircraft development plan to formulate a detailed progress schedule of aircraft safety activities and define the activities at every node.

The principle of safety activities are as follows: developing top-level safety requirements that should be integrated into aircraft process schedule; coordinating the activities related to certification, reliability, maintenance and quality assurance; defining the duties and responsibilities of safety activities to perform and review safety plan easily.

The safety plan is to:

- identify the safety requirements of aircraft level, including duties and responsibilities of safety design and analysis;
- determine the applicable safety standards;
- identify project safety organizations and specify the responsibilities of the internal organization and relations between the partners and/or suppliers related to the safety process;
- describe safety activities and deliverables;
- specify the critical project milestones that need to be completed;
- management principle of safety requirements verification and validation;
- Relations with other plans (e.g., certification plan, verification and validation plan, process assurance plan, etc.).

2.3.2 The Safety Activities and Responsibilities of Related Stakeholders

2.3.2.1 The Safety Activities and Responsibilities of Aircraft Manufacturer

The safety activities and responsibilities of aircraft manufacturer consist of two sections: the safety activities of aircraft manufacturer and the information provided by aircraft manufacturer for suppliers.

2.3.2.1.1 The Safety Activities of Aircraft Manufacturer

The safety activities of aircraft manufacturer are divided into two levels: the aircraft level safety activities and the system level safety activities.

1. Aircraft level safety activities

The responsibilities of aircraft level safety activities include the following aspects:

- preparing and assigning the safety activities of the aircraft project;
- coordination and information exchange among each safety working group;
- formulating the safety plan;
- formulating safety principles and ensuring them to be carried out through the results of review and validation/verification activities;

- ensuring the safety objectives of aircraft level to be met;
- aircraft level safety synthesis.

    The main purposes of the aircraft level safety activities:

- formulating the management principles of the safety activities as well as defining safety common data to make the decomposed activities at each level clearly and orderly;
- ensuring the implementation of safety activities to comply with the safety plan;
- integrating all the results of safety activities and showing the aircraft to meet the safety objectives.

    The aircraft level safety activities include three aspects listed below:

a. The formulation of management principles and methods of aircraft level safety activities

    This includes the following aspects:

- identifying aircraft level safety requirements and safety philosophy of;
- specifying the methods, contents and forms of safety assessment;
- specifying the methods, contents and forms of safety synthesis activities;
- specifying the methods of establishing the common database and conducting quantitative calculations;
- specifying the method of using empirical data.

b. The technical activities of aircraft level safety

    The technical activities of aircraft level safety include the following aspects:

- AFHA
- PASA/ASA
- ZSA
- PRA
- Aircraft level CMA

c. Aircraft significant item list and associated monitoring

- The coordination and integration of aircraft level safety activities
- The integration activities
- Continuous airworthiness monitoring
- The safety common data of aircraft
- The collection and usage of empirical data
- The interface activities with other activities and external systems of the aircraft

2. System level safety activities

For the aircraft manufacturer, the system level safety activities are the continuation of the aircraft level safety activities.

The safety engineers of system level should implement the system level activities according to the safety management principles and methods issued by the working group of aircraft level safety.

The safety assessment activities required by related systems are performed to demonstrate that system level safety objectives and requirements are met, and issue the safety requirements to the following objects:

* the implementation of other system functions;
* the assembly activities;
* the suppliers.

Based on the results of safety analyses provided by the suppliers, such as PSSA, SSA, and CMA reports, the safety engineers of aircraft manufacturer perform the review, validation, and verification for them. Then incorporate the results of safety activities into the system level safety assessments of the aircraft manufacturer, such as issuing their own PSSA, SSA, CMA, and other safety assessment reports. The PSSA, SSA, and CMA reports used for certification should be accomplished, or at least issued by aircraft manufacturer.

The results of system level safety assessments will be transferred to other related systems and integrated into their activities of design and safety assessments.

After the validation and verification for the results of system level safety assessments have completed, such results are transferred to the working group of aircraft level safety to be used for the aircraft level assessment and synthesis.

For parts of the system level safety activities of the aircraft manufacturer, such as system level safety plan and SFHA, may be accomplished by the system suppliers.

2.3.2.1.2 Information Provided by Aircraft Manufacturer for System Suppliers

The aircraft manufacturer should provide the following information for system suppliers as the inputs of their safety activities:

* System level PTS;
* AFHA report;
* PASA report;

- safety common data, such as flight phases, risk time, classification of failure condition, and probability of external events;
- safety principle, such as the format of safety documents as well as guidance and principles of safety activities;
- SFHA report (SFHA may also be accomplished by the system suppliers as appropriate);
- safety plan at system level (This plan may also be accomplished by the system suppliers as appropriate.).

2.3.2.2 Safety Activities and Responsibilities of Suppliers

The safety activities of supplier's aim at demonstrating that the safety requirements identified in the system PTS are satisfied. Responsibilities of the competent department for safety of suppliers are as follows:

- communicating, connecting, and coordinating with the competent department for safety of aircraft manufacturer;
- accomplishing the technical activities of system level safety, such as SFHA (if appropriate), PSSA, SSA, and CMA;
- formulating safety philosophy and ensuring it to be carried out through the review and validation/verification activities;
- Management of the safety activities, such as safety requirements, DALs, and validation/verification activities;
- ensuring the system level safety requirements can be met.

According to the contents and depth of safety activities required by the system PTS, suppliers should accomplish the following activities: formulating the safety plan, accomplishing the SFHA (if appropriate), PSSA, SSA, CMA, and other activities, and submitting the interface data generated in the safety process to the competent department for safety of aircraft manufacturer.

2.3.3 Other Activities Related to Safety Management

2.3.3.1 Safety Training

The safety training plans should be formulated according to the safety management and control procedures as well as safety plans, and then the training on safety assessment process and analysis tools, safety requirements validation and verification, and other aspects should be implemented to ensure the accomplishment of the safety activities.

Personnel accepting the safety training in the development departments mainly involve the safety personnel, design personnel, test personnel, and so on. According to the tasks, technical difficulties and levels of the training, along with different technical levels and objects of safety training, the safety training plans are formulated and associated activities are carried out.

2.3.3.2 Validation and Verification of Safety Requirements and Associated Management

A validation/verification process should be included in the safety assessment process. It is in relation with the overall validation/verification plans at aircraft and system level.

The assumptions, which are used and generated in the process of safety requirement analyses, function definitions, definition of failure condition classifications, and safety assessments, should be managed and controlled effectively, and their validation plans should be formulated. The completeness, correctness, timeliness, and traceability should be ensured during the management and transmission of assumptions.

2.3.3.3 Safety Review

The safety review refers to the internal review of the applicant, but not the review of the certification authority. The scheduled safety reviews are conducted in accordance with the safety plan in order to ensure the safety activities to be carried out timely and orderly, and the quantitative and qualitative safety requirements are satisfied. All the data of safety reviews should be filed timely to ensure the traceability of safety reviews and timeliness of relevant safety activities.

In terms of the analyses, reviews, and assessments on design, concepts, methods and results, the safety design is reviewed whether it conforms to the safety plan, airworthiness requirements, design specifications, etc. Therefore, it can be timely to discover the potential design deficiencies, identify the high-risk areas of safety, put forward some suggestions for improvement, and effectively take corrective actions.

The safety reviews can be divided into preliminary design review, critical design review, phase design review, and final design review, which can be conducted in combination with the reviews on the performance, reliability, maintainability, and other aspects of the aircraft.

2.3.3.4 Management of Safety Information

The safety information here mainly includes but not limited to the following:

• safety information of system or equipment suppliers;

- safety information of market researches and statistics;
- safety information of design, manufacturing, final assembly, testing, and post-operation;
- safety information required by the development and its alteration, adding or deleting.

It is required to establish the closed-loop system of safety information to ensure the collection, transmission, feedback, analysis, disposition, management, and other activities of safety information to be carried out smoothly. The basic principle of safety information management is to maintain the traceability of safety common data used during safety assessments process and its consistency with the aircraft characteristic data (such as the mission/flight envelope, failure rate, probability of external events, at risk time, and failure condition classifications, etc.).

2.3.4 An Example of Safety Functional Organization

At the beginning of the aircraft development, it is necessary to establish a sound safety functional organization to determine the duties and responsibilities related to safety such as safety organization, the design department, manufacturing, assembly, and other departments. The safety organizations are responsible for the following safety activities:

- review and decision-making of safety activities;
- aircraft level safety activities;
- system level safety activities;
- assembly/test safety activities;
- safety activities of suppliers;
- continuous safety monitoring activities.

For a particular aircraft project, the safety activities are as shown in Fig. 2.3.

The executive agencies of safety management can be divided into seven parts, including:

Group A—Safety assessment committee

The safety assessment committee, an organization supporting the decisions of major safety issues, is composed of design personnel and experts in the industry and is in the charge of the chief designer of the project.

The safety assessment committee, as the origination supporting the decisions of safety activities, supervises and instructs the activities of every other safety department. The contents are as follows:

1. reviewing the safety plan and verification plan;

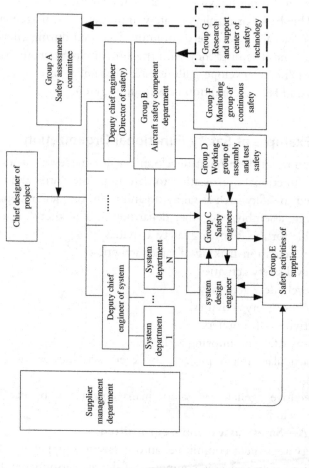

Figure 2.3 Typical safety functional organization.

2. reviewing the activities implemented in every development phase:
- aircraft level FHA;
- FHA and PSSA of critical systems
- special topics of major CCA;
- assessing major technical problems of safety;
- assessing the safety design and verification activities at aircraft, system, equipment (suppliers), assembly and operational level.

According to the progress in the development phase, the safety assessment committee organizes regular and irregular review meetings. The review comments made by the committee are submitted to the leaders and relevant departments as the reference for the decisions made by competent department. It is required for the committee to ensure the close connections between the safety assessment and system design throughputting forward guidance and suggestions to the designers.

Group B—Competent department of aircraft level safety

The competent department of aircraft level safety is the major management and implementation department of aircraft safety activities. Under the department, there are working group of assembly/test safety and monitoring group of continuous safety. Meanwhile, it leads the support center of safety technology research in operation. At the aircraft level, it is mainly responsible for:
- preparing and assigning the safety activities of every department;
- the coordination and information exchange among each safety working group, and formulating the safety plan;
- ensuring the implementation of safety policies;
- ensuring the application of safety policies through the results of review and validation/verification activities;
- ensuring aircraft level safety objectives to be met;
- aircraft level safety syntheses.

In addition, the competent department of aircraft level safety is responsible for ensuring that all training courses required by the safety activities have the conditions for implementation, and that safety engineers can possess a necessary level of expertise by these courses. In this book, we define safety engineers as those system design engineers who possess expertise in safety, and safety specialists as those who offer guidance and advisory on safety/reliability.

Group C—Working group of system level safety

The working group of system level safety is mainly composed of the safety engineers of each system and is in the charge of system department.

It closely cooperates with the design teams but is independent of the design personnel, and it is instructed by the competent department of aircraft level safety. In the event of contradictions, the working group need to report to the vice chief engineer of safety and the director of department directly. The qualification of the safety engineers is approved by Group A.

In order to guarantee the cooperation between safety personnel and system design personnel, all safety technical documents published at the system level are jointly signed and issued by both sides (safety and design).

For some specific project, safety assessments may not be conducted by the system level safety group. They can be outsourced. However, in any case, the system safety group should be responsible for the validation/verification activities and deliverables.

Group D—Working group of assembly/test safety

The working group of assembly/test safety is responsible for the safety activities of final assembly of aircraft, ground test, flight test, as well as operation task and support. The Group D mainly ensures the safety requirements put forward by the aircraft level and system level safety working groups to be implemented and verified, including:

- confirming the assumptions and results of safety analyses through tests;
- incorporating the maintenance tasks related to safety analyses into the operation documents and product support documents;
- incorporating the requirements generated by safety analyses into the installation and integration activities;
- incorporating the procedures and tests determined by safety analyses and assessments into the operation manual (flight crew manual);
- incorporating the results of safety proposals into the Master Minimum Equipment List.

Group E—Safety department of suppliers

The safety activities of system, subsystem, or equipment suppliers aim at demonstrating that the safety requirements put forward by the system level working group or in the PTS have been satisfied. The safety activities of system level are responsible for supervising and inspecting that of supplier level.

Group F—Monitoring group of continuous safety

The activities of monitoring group of continuous safety are to ensure the safety monitoring in the aircraft operation for updating, correcting safety data (such as failure rate, Mean Time Between Failure (MTBF), and probability of events), and recording the events occurred in service.

Group G—Research and support center of safety technology

The research and support center of safety technology is not a permanent entity, but a technology-supported body specific to the actual demand of projects, which is formed by organizing the professionals of domestic industry. It is mainly responsible for studying the technologies, methods, and tools required by the safety assessment at each level. Group G is instructed by Group A and Group B in terms of operation.

2.4 SAFETY DOCUMENTS OF AIRCRAFT DEVELOP LIFE CYCLE

During the aircraft development process, a variety of safety documents will be produced to ensure the traceability. The main safety documents include the top-level documents, methodology documents, and technical documents. The top-level documents include documents such as safety program and safety plan. The methodology documents are guidelines for various safety assessment processes and methods such as AFHA, FTA, and CMA, and so on. The technical documents include a variety of safety technical documents such as AFHA document, PSSA document of a specific system, and FMEA document of certain equipment. In addition, the following documents will be produced during the safety assessment process: the periodic summary documents, review documents, validation/verification documents, as well as the technical documents related to safety, such as CCMR document.

A part of above-mentioned safety documents will serve as the verification documents for the certification authority to support the aircraft certification. The aircraft level safety documents include the aircraft safety plan, the AFHA report, the PASA report, and the ASA report, and so on. The system level safety documents include SFHA reports, system PSSA reports, system SSA reports, and system FMEA reports, and so on. The equipment level documents include the equipment level FMEA reports. At the same time, all the three-level activities should complete the ZSA, PRA, and CMA reports for CCA.

2.5 ADDITIONAL TOPICS RELATED TO SYSTEM SAFETY ASSESSMENT OF CIVIL AIRCRAFT

There are also some additional contents related to safety assessment of civil aircraft, such as the CCMR, MMEL, and Extended Operations (ETOPS).

These contents, as part of the assurance for the civil aircraft safety, are related and complementary to the process of system safety assessment.

2.5.1 Certification Maintenance Requirement

2.5.1.1 An Overview of Certification Maintenance Requirement

When safety assessment is conducted for transport category airplanes, the failure detection requirements are put forward to avoid the occurrence of hazardous and catastrophic failure conditions under §25.1309. These requirements are further reviewed and identified as the CMR which serves as the important contents for obtaining the airworthiness certificate and published to the user for proper execution and monitoring.

A CMR is a required scheduled maintenance task established during the design and certification process of the aircraft as an operating limitation of the TC or Supplemental Type Certificate (STC). The CMRs are a subset of the instructions for continued airworthiness documents identified during the type certification process. A CMR usually results from a formal, numerical analysis conducted to show compliance with the catastrophic and hazardous failure conditions [3].

The CMRs are intended to detect latent failures that have significant effects on aircraft safety. Failures of the CMRs will not have catastrophic effects on the aircraft, but the combination of this very type of latent failure with one or more other specific failures or events will result in a catastrophic or hazardous failure condition. A CMR can also be used to establish a required task to detect an impending wear-out of an item whose failure is associated with a hazardous or catastrophic failure condition.

Although the CMR tasks and inspection intervals are obtained through a series of analyses, the purpose and analysis methods of the CMR are quite different from the MSG-3 associated with Maintenance Review Board activities. Although both types of analyses may produce equivalent maintenance tasks and intervals, it is not always appropriate to substitute a CMR with an MSG-3 task.

The CMRs are necessary to meet the requirements of the safety assessment under §25.1309 and should not be confused with the required structural inspection program that are developed by the TC applicant for meeting the requirements of damage tolerance. The CMRs are to be developed and managed separately from any structural inspections programs.

In a quantitative calculation demonstrating compliance with §25.1309, exposure time will have a significant effect on the failure probability of failure condition since the exposure time is a critical factor for latent failure. The intervals for CMR tasks should be designated in terms of flight hours, cycles, or calendar time, as appropriate.

2.5.1.2 Identification of Candidate Certification Maintenance Requirements
2.5.1.2.1 Principles for identification of Candidate Certification Maintenance Requirements

The CCMRs generally focuses on significant latent failure that can cause catastrophic or hazardous failure conditions to the aircraft if combined with one or more additional specific failures or events.

2.5.1.2.2 Method for identification of Candidate Certification Maintenance Requirements

The CCMRs are usually derived from formal and quantitative safety analyses, which can determine whether there are regular task requirements to demonstrate the compliance with §25.1309. The SSA should identify as CCMRs the maintenance tasks intended to detect latent failures that would, in combination with one or more specified failures or events, lead to a hazardous or catastrophic failure condition.

In addition, other tasks that are not generated from a formal safety assessment but are based on the properly certified reasonable engineering judgments can also serve as the CCMRs. Such a judgment should include the logic for verifying that it is a CCMR and the data and experience supporting the logic. With reference to AC 25-19A, CCMRs may also be identified for latent failures that would, in combination with one or more specified failures or events, lead to a major failure condition that is not identified and assigned a task via the MSG-3 process (however experience has shown that these cases are rare).

2.5.1.3 Selection of Certification Maintenance Requirements

Each CCMR should be reviewed by the Certification Maintenance Coordination Committee (CMCC) and a determination made as to whether it should be a CMR. The applicant should provide sufficient information to the CMCC to enable an understanding of the failure conditions and the failure or event combinations that result in the CCMR. CCMRs are evaluated in the context of failure conditions in which they are involved, e.g., whether the latent failure is part of a dual failure, or a more severe failure condition.

The CMR designation should be applied in the case of catastrophic dual failures where one failure is latent. The CMR designation should also be applied to tasks that address wear out of a component involved in a catastrophic failure condition that results from two failures. The interval for the CMR tasks should be chosen such that the system safety analysis assumptions are protected in service, while allowing flexibility for the airplane operators to manage their maintenance programs. In the case where the system safety analysis does not specify an interval, the CMCC may establish an interval that is less than the life of the airplane considering factors that influence the outcome of the failure condition, such as the nature of the fault, field experience, or the task characteristics.

The CMR designation may not be necessary if there is an equivalent MSG-3 task, or an approved Aircraft Flight Manual procedure, to accommodate the CCMR.

2.5.2 Master Minimum Equipment List

Modern aircraft adopts the design of high reliability and redundancy. The TC proves that the aircraft is safe and reliable when all equipment work properly. However, failures may still occur and can cause flight delays or cancelations, increasing the operator's operating costs. Therefore, in the presence of inoperative equipment in the aircraft operation process, balancing the acceptable level of safety and airline revenue properly is the reason and objective for proposing and formulating the MMEL.

The effects on safe operation are different for different airborne equipment failures. For example, the failure of a reading light in the cabin obviously will not affect the safe operation of the aircraft. When an engine of the aircraft fails, no operator dare to allow such an aircraft to continue flying. However, during the actual use, it is often difficult for the operator to determine whether the failure of a certain item will affect the safety of the aircraft. For example, there are three separate hydraulic systems on board, such that when the indicator of the second hydraulic system fails, it is difficult for the operator to determine how much the failure will affect the safe operation. Thus, operators need an efficient and reliable reference document to quickly determine whether the failure will affect flight safety, and whether the aircraft can be released in the case of failure. The MMEL/Minimum Equipment List (MEL) are the reference documents required by the operator, and they can assist the operator in determining whether to release the aircraft or not.

The MMEL must meet the relevant regulations and should be approved and accepted by the Administration. The MMEL/MEL does not include those items that significantly affect the safety (e.g., inoperative engine, rudder and aileron, elevator malfunction, nonretractable landing gear, etc.) and that do not significantly affect the safety (e.g., kitchen equipment, etc.). The MMEL/MEL may be considered as an approved deviation from the certified type design.

2.5.2.1 Overview

The MMEL is developed by the airworthiness authorities of the aircraft manufacturing country, to guide the aircraft users and airlines to compile the programmatic document of MEL specifically. It specifies the approved inoperative instruments and equipment with which this type of aircraft can be released, and it also has principled requirements on the minimum number of operating instruments and equipment for the release as well as on the restrictive clause for the release with failures retained.

The MEL is an important technical document formulated by the aircraft operator and approved by the national airworthiness authority. The formulation of MEL is based on the MMEL. MEL is a document developed to allow aircraft of specific type and with serial numbers and registration numbers to be released with inoperative equipment and systems within a certain period, the formulation of which is based on the MMEL according to the differences of aircraft configurations selected by the airlines, and it is combined with the company's operating ability, experience, and other differences.

The MEL shall comply with or be stricter than the MMEL for the relevant aircraft type. The main purpose of the MEL is to make full use of the safety margin of aircraft design and to allow an aircraft with failures to continue to fly within the prescribed period in the context of ensuring the safe operation. The reasonable use of MEL can effectively improve the utilization rate of aircraft and flight punctuality as well as reduce the operating costs. It is noteworthy that the MEL is not the maintenance standard of the aircraft, and a flight with failures is not advocated at all.

The maintenance department shall complete the troubleshooting activities as soon as possible.

The MEL is similar to the MMEL. Their difference is that the MEL is formulated for a particular operator and a certain aircraft or a few aircraft, whereas the MMEL is formulated for all aircrafts of this type. The MEL of the operator shall be based on the MMEL of a specific aircraft type and model approved by the authorities. The MEL is stricter than the approved MMEL.

2.5.2.2 Formulation Method of the Master Minimum Equipment List
2.5.2.2.1 Qualitative Analysis

When formulating an MMEL, a qualitative analysis shall be performed first and all factors affecting the aircraft operation shall be taken into full consideration. These factors include functional transformation of spare components, adjustments to operational limits, adjustments to flight crew and maintenance procedures, and adjustments to the minimization of the workload of the flight crew.

Qualitative analysis is based on engineering judgment. These analyses may be based on previous experience gained from the MMEL release scenarios. However, in regard to the same items for different types of aircraft, qualitative analysis may not be directly adopted. The differences of the aircraft architecture as well as system operation and use shall be taken into consideration.

A complete analysis of the interactions between systems can ensure that multiple failures will not lead to unsatisfactory safety levels. In addition, flight test and simulated test can help those personnel who formulate the MMEL to assess the effects of inoperative equipment on aircraft safety and the workload of the crew.

When equipment of critical systems is included in the MMEL, their failure characteristics shall be taken into account in the safety assessment process. For an item with temporarily inoperative equipment, additional risks arising from its flight shall be assessed. The occurrence probability of risks should comply with that at the acceptable level determined during the type certification process.

When the consequences of system failures are proven to be at the secondary level, the next step is to perform a qualitative analysis of the next level of critical failure. When a failure occurs, it will have a very negative effect on the operation of the aircraft in combination with other specific failures. Then this failure will be labeled "the next level of critical failure."

For example, if the low pressure valve of the engine fails, the aircraft will not be released. If the valve failure is in the "off" position, the engine will not start; if the valve failure is in the "on" position, the affected engine will not be isolated in the case of engine fire.

Therefore, when we perform qualitative analysis, it is necessary to analyze whether the next level of critical failure will lead to hazardous effects on the flight safety after the determination of the failure of candidate MMEL items.

Qualitative Safety Analysis: If equipment is included in the MMEL, a qualitative analysis should be made to determine the effects of inoperative equipment on all other aspects of aircraft operation. The qualitative analyses must consider its effects on the workload of the crew, on multiple equipment of MMEL, as well as on the complexity of maintenance and operation procedures. In addition, the qualitative analysis can also reflect relevant experiences gained from operations with the use of MMEL.

Redundancy Analysis: If the function of the selected component or system can be substituted by equipment, then the component or system can be considered as the redundant item in the condition that the alternative equipment of the equipment is verified to work properly. Redundancy cannot be act as a good reason for categorizing the equipment into the MMEL if the basis of aircraft type certification requires two (or more) functions or information sources. In this case, another verification method can be adopted, such as safety analysis.

These analyses are carried out to demonstrate that even if an item fails, the safety objectives have already been met and the aircraft can be safely operated.

2.5.2.2.2 Quantitative Analysis

Through the qualitative analysis described above, aircraft manufacturers can formulate the MMEL items. However, the rectification interval of each item shall be determined by the quantitative analysis.

In the process of formulating the MMEL, the quantitative analysis mainly adopts the method of SSA, and the rectification interval of each MMEL item is worked out through the FTA.

2.5.3 Extended Operations

ETOPS is the abbreviation of "Extended Operations." ETOPS is a specific requirement raised by the authorities to ensure the continuous safe flight of the aircraft. In the condition of one engine failure or critical system failure of two-engine or multi-engine aircraft, it is required to safely land in the alternate airport according to the ETOPS rules with the remaining operative engine(s). The ETOPS with a higher ability and longer time means that airlines can use two-engine or multi-engine aircraft to perform more non-stop transoceanic tasks, as it can shorten the approved routes and reduce costs.

Airplane system safety assessments for ETOPS are addressed under the specific objectives of FAR25.901(c) and 25.1309, considering the maximum flight time and longest diversion time for which the applicant seeks approval. The main impact that the ETOPS will have on airplane system safety assessments is a potentially more severe failure condition when considering the long-range and maximum ETOPS diversion distances associated with a maximum ETOPS flight. For example, a failure in an airplane's environmental control system resulting in either a very hot or a very cold cabin temperature could be potentially life-threatening during a 5-hour diversion, whereas the same failure would merely be an uncomfortable inconvenience during a 30-minute diversion. What may be considered a minor or major effect during a short diversion may have a hazardous or even catastrophic effect over a longer period. Such time-related effects must be considered in the safety assessments of these types of failures to ensure that any potentially unsafe failure conditions are identified and the proper classification is defined.

Section K25.1.1 of Part-25 Appendix K, *The airplane-engine combination must comply with the requirements of part 25 considering the maximum flight time and the longest diversion time for which the applicant seeks approval*, it requires the applicant to show that the airplane systems meet the safety objectives of FAR25.901(c) and 25.1309 for any failure condition that has a more severe failure effect when considering a maximum ETOPS diversion following the failure.

REFERENCES

[1] SAE ARP 4754A. Guidelines for development of civil aircraft and systems. SAE International; 2010.
[2] SAE ARP 4761. Guidelines and methods for conducting the safety assessment process on civil airborne systems and equipment. SAE International; 1996.
[3] FAA AC 25.1529-1A. Instructions for continued airworthiness of structural repairs on transport airplanes; 2007.

CHAPTER 3

Aircraft Functional Hazard Assessment

Content

3.1 CONCEPT

The civil aviation has developed a relatively mature and improved safety assessment method. Thus, *SAE ARP 4761* was produced for providing guideline to assess system safety of civil aircraft [1]. Generally, the process of safety assessment is not a one-time task, just like the Functional Hazard Assessment (FHA) that we introduced in this chapter.

What is FHA?

The safety requirements to be captured at the beginning of aircraft design are used as inputs to subsequent safety design. FHA can usually be expressed as a systematic and comprehensive assessment method which is used to examine the aircraft or system functions to identify possible failure conditions and classify them according to the severity of their effects. FHA should be implemented at the early stage of the development process after determining aircraft functions. The input data of FHA should be updated when new functions or failure conditions are identified or errors are found in subsequent design activities. Moreover, combined

Civil Aircraft Electrical Power System Safety Assessment
DOI: http://dx.doi.org/10.1016/B978-0-08-100721-1.00003-0

with design data, the safety objectives established by FHA should be continuously updated until the development process is completed, to avoid significant omissions or errors at later stages. Therefore, FHA is essentially an iterative process, i.e., FHA report is a dynamic document that is used throughout the development cycle.

Usually, FHA can be divided into Aircraft Functional Hazard Assessment (AFHA) and System Functional Hazard Assessment (SFHA). AFHA regards the overall aircraft as the object to analyze the failure conditions that may affect continued safe flight and landing of the aircraft in its whole flight envelope and different flight phases. AFHA adopts a top-down functional analysis method to identify failure conditions, assess their effects and classifications, and determine the safety objectives of system design.

AFHA is the starting point of the capture and allocation of aircraft safety requirements, providing inputs for Preliminary Aircraft Safety Assessment (PASA). It is connected to SFHA through PASA and aircraft level requirement documents. The results of FHA and PASA are the inputs to Preliminary System Safety Assessment (PSSA) and System Safety Assessment (SSA).

AFHA and SFHA are performed in a similar way. The difference is that AFHA considers the overall aircraft as the object, starting at the aircraft concept development stage, and SFHA takes the systems as the object, starting at the preliminary system design stage. For more information on the detailed process of FHA, please see the relevant contents in Chapter 4, System Functional Hazard Assessment. This chapter only describes the characteristics of the AFHA process.

3.2 AIRCRAFT FUNCTIONAL HAZARD ASSESSMENT INPUTS

The first step for implementing AFHA is to obtain the necessary data, which usually include several aspects that are described below:

3.2.1 The Aircraft Level Function List

The aircraft level function list should specify the aircraft level functions and interactive functions among them, the possible operating mode of each function, the complexity and novelty of functions, the limits of functions on aircraft operation, and so on. In addition, the aircraft systems performing these functions, along with the interfaces between each other, should be described.

The determination of aircraft level functions usually based on engineering judgement and experience. In general, typical aircraft level functions include "Ground Control", "Flight Control" (i.e., "Roll Control", "Yaw Control", and "Pitch Control"), "Environment Control", "Thrust Control", "Provide Communications", "Provide Protections", "Provide Consumables and Power", and so on. The detailed list is presented in Table 3.1.

As an example, the "Roll Control" function is associated with flaps, ailerons, spoiler, engine, and autopilot. The associated failures will cause a loss or significant effect of the "Roll Control" function, such as "Unprotected Flap Asymmetry beyond Limits", "Loss in Ailerons during Rolling Maneuvers", and "Unprotected Aileron Servo Handover".

3.2.2 The Aircraft Design Objectives and Users' Requirements

The aircraft design objectives should specify the overall design requirements of the aircraft, mainly including the average flight duration, seat occupancy number, range, designed speed, flight altitude, field performance, service life, temperature of the service environment, average maintenance interval, operating conditions (e.g., icing, etc.), and so on. In addition, requirements of regulation (i.e., airworthiness and operational) and users should be included in design objectives.

3.2.3 Main Features of the Aircraft

It should specify the main features of the aircraft, such as the general layout, number of engines, means of providing power for systems (e.g., fly-by-wire, hydraulic, etc.), and avionics system with or without IMA.

3.3 PROCESS OF AIRCRAFT FUNCTIONAL HAZARD ASSESSMENT

Usually, the process and characteristics of AFHA should include the following aspects:

3.3.1 Review and Confirm Aircraft Level Functions

The aircraft level function list is an input to AFHA. In other words, AFHA process does not define aircraft functions, but reviews and confirms whether the aircraft functions are appropriate for use. Aircraft level functions are generally divided into internal and external functions.

Table 3.1 An example list of aircraft level functions

Aircraft functions	Aircraft subfunctions	Associated systems functions
Ground Control	Speed Control on Ground	Ground Spoilers Control Wheels Braking Control Thrust Reverser Control
	Direction Control on Ground	Rudder Control Nose Wheel Steering Control Differential Braking Control Differential Thrust Control
	. . .	. . .
Flight Control	Roll Control	Roll Axis Control Roll Trim Control
	Yaw Control	Yaw Axis Control Yaw Trim Control
	Pitch Control	Pitch Axis Control Pitch Trim Control
	. . .	. . .
Environment Control	Atmosphere Control	Pressure, Temperature, Humidity, and Ventilation Control
	Provide Life Support and Comfort	Provide Crew Satisfaction and Comfort
	Provide Lighting	Provide Interior Lighting Provide Exterior Lighting
	. . .	. . .
Provide Thrust and Control	Thrust Generation Thrust Control	Engine Starting Control Engine Thrust Control Automatic Thrust Control
	. . .	. . .
Provide Communications	Provide Interior Communications	Provide Crews Communications
	Provide Exterior Communications	Provide Communications with Ground Control Provide Communications with Other Aircraft
	. . .	. . .

(Continued)

Table 3.1 (Continued)

Aircraft functions	Aircraft subfunctions	Associated systems functions
Provide Protections Against Natural and Induced Environment	Provide Protections Against Environmental Hazards	Lightning and Electromagnetic Protection Single Event Upset Protection Ice, Rain, and Defogging Protection Anticollision Protection Wind Shear and Gust Protection Protection Against Threat
	Provide Protections Against Intrinsic Hazards	Engine Burst Protection Depressurization Protection Fire Protection
	. . .	. . .
Provide consumables and power	Provide Consumables	Provide Fuel Provide Water Provide Oxygen
	Provide Power	Provide Hydraulic Power Provide Electrical Power Provide Pneumatic Power
	. . .	. . .
. . .	. . .	. . .

Internal functions refer to the main functions of the aircraft and the interface functions among its systems. External functions refer to the interface functions between aircraft and other aircraft or a ground system. An aircraft level function list should widely draw on the previous engineering experience of similar aircraft and should be made combined with the aircraft design requirements. Meanwhile, the opinions and suggestions of experts from various engineering development departments, suppliers, authorities, and users must be considered.

Analysis of aircraft level functions is conducted layer by layer to determine all the possible functions and subfunctions in any operating mode and condition. Usually, it only analyzes the functions, without implementing associated system, equipment, or architecture. The aircraft level function list should be performed layer by layer, so that lower level functions can be determined according to the upper level functions. To ensure the completeness and correctness of the allocation process of aircraft level functions as well as the traceability between aircraft level functions and system level functions, the lowest aircraft level functions (e.g., "Provide Hydraulic Power", "Provide Electrical Power", etc.) can be directly related to a specific system (It is suggested that each lowest aircraft level function only corresponds to one aircraft system.), which depends on analysts' familiarity with the main features and the development progress of the aircraft.

> Note: Aircraft level function list should be updated timely as the update of input data of AFHA.

3.3.2 Identify and Describe Failure Conditions

One of the most important tasks of AFHA is to identify and describe the failure conditions of each aircraft level function. When identifying the failure conditions of functions, full consideration should be taken into the internal functions, external functions, and each environment condition/ event and emergency configuration of the object in the aircraft level function list. In addition, combined function failures should also be considered to determine whether there is a more severe failure condition. The analysis of combined function failures is relatively complex and requires comprehensive understanding of the internal and external functions of the aircraft. Combined function failures can generally be considered from the following aspects:

- Failures related to environmental events and emergency configurations (e.g., weather, ditching, engine out, fire, decompression, etc.);
- Failures related to other failures (e.g., main function failure in case of alerting or protection function failure, etc.);
- Failures related to specific flight phase (e.g., takeoff, landing, etc.).

3.3.3 Determine the Effects of Failure Conditions

The operational condition or flight phase of failure conditions should then be determined. In some cases, the effect of the same failure condition is different in different flight phases. Thus, the different flight phases and their effects should be listed. For each failure, possible scenarios that describe the effects associated with the envisaged failure should be considered. The different combinations in the scenarios are mainly derived from the scenarios that describe the effects in the case of a single function failure, considering all the possible additional failures or events. Particularly, different failures that can lead to the identical effects on aircraft in different flight phases should be combined into the same failure condition with a unique reference and title.

3.3.3.1 The Effects of Failure Conditions on Aircraft or Personnel

The effects of each functional failure condition on aircraft or personnel (i.e., flight crew and occupants) must be determined when conducting AFHA. Usually, the effects can be identified by consulting experienced experts. To accurately analyze the effects of failure conditions on aircraft or personnel, various factors must be comprehensively taken into consideration, including the flight phase, failure occurrence indication to crew, abnormal weather conditions (e.g., ice, snow, etc.), precautions taken by the flight crew in response to failures, and excessive workload of the flight crew caused by dealing with failures. Moreover, when assessing the effects of failure conditions, potential factors that may affect the flight crew while dealing with emergency situation (e.g., smoke, communication interruption, cabin pressurization interference, etc.) should be taken into consideration.

In addition, whether the effects of failure conditions in different flight phases are identical should be analyzed. Furthermore, the respective effects on personnel, aircraft control, and processing requirements should be assessed. Moreover, the following principles should be followed when analyzing the effects:

- The effects of indicator system errors are usually more severe than the loss of indicator system.
- The requirements of the aircraft for the pilots should be understood and defined, including the operational requirements in each flight phase, in order to analyze the requirements and effects of failure conditions when it is operated by the pilots.
- The capacity of pilots to deal with the failure conditions should be based on the requirements of the aircraft for pilots, and the capacity of an individual pilot cannot be served as the basis for analyzing the effects of failure conditions.
- It is required to define whether the annunciated failure conditions are identical to the unannunciated failure conditions.

- Necessary analyses are required to indicate the credibility and correctness of their results.

3.3.3.2 Determine the Classifications of Failure Conditions
The classification of failure conditions should be determined according to the severity of their effects on the aircraft, flight crew, and occupants. According to AC25.1309-1B (Arsenal) [2], the classification of failure conditions of civil aviation aircraft can be divided into five classifications, namely, Catastrophic, Hazardous, Major, Minor, and No Safety Effect.

3.4 AIRCRAFT FUNCTIONAL HAZARD ASSESSMENT OF TRANSPORT AIRCRAFT

This book takes a two engine transport aircraft as an example. The aircraft is designed to carry 75 passengers with an average cruising speed of 0.78 Mach, a maximum range of 1200 nautical miles, an average flight duration of 1.2 hours, and a 50,000-hour service life.

An example of some aircraft level functions is listed in Table 3.1.

The whole book takes the Electrical Power System (EPS) as a case study, and in this section, the failure conditions identified in AFHA are analyzed. This case is simplified, and due to the differences in specific projects, the analysis and the results of this AFHA case are not necessarily applicable to other AFHAs. Therefore, this case is for reference only. The details regarding the identification and combination process of failure conditions are described in Chapter 4, System Functional Hazard Assessment. The worksheet example for an EPS AFHA is shown in Table 3.2.

According to the results of AFHA analyses, all of the failure conditions of catastrophic and hazardous are listed in Table 3.3.

3.5 PRELIMINARY AIRCRAFT SAFETY ASSESSMENT

The purpose of PASA is to identify the interactions and dependencies between aircraft systems, to assess the aircraft level failure conditions identified by AFHA, to evaluate the aircraft architecture, and to determine whether the identified aircraft level safety requirements can be met.

PASA should be started during the initial aircraft architecture development phase, and its process is iterative throughout the whole development process.

At the beginning of the PASA process, input data should be gathered; these data constitute the minimum data necessary for a safety analyst to

Table 3.2 Worksheet example for the EPS in AFHA

Function	FC	FC REF.	Flight phase (see Section 4.5.3.2)	Effect of the FC on aircraft or personnel 1. Aircraft 2. Flight crew 3. Occupants	Classification	Verification methods	Reference to supporting material	Remarks
Provide Electrical Power	Total Loss of AC Network	24–FC–1	T F1–F4 L	Aircraft: It causes the loss of all AC electrical equipment and prevents the aircraft to continue safe flight or landing. Flight crew: It causes their disability to control the aircraft. Occupants: It may cause casualties of all occupants.	Catastrophic	Qualitative and quantitative FMEA, Qualitative and quantitative FTA, CCA	Note 1	
	Loss of Normal AC Network	24–FC–2	T F1–F4 L	Aircraft: Part of AC electrical equipment cannot work, greatly reducing the safety margin of aircraft power supply. Flight crew: It greatly increases the workload of the crew. Occupants: No effect.	Major	Qualitative and quantitative FMEA, Qualitative and quantitative FTA, CCA		Note 2

(Continued)

Table 3.2 (Continued)

Function	FC	FC REF.	Flight phase (see Section 4.5.3.2)	Effect of the FC on aircraft or personnel 1. Aircraft 2. Flight crew 3. Occupants	Classification	Verification methods	Reference to supporting material	Remarks
	Total Loss of DC Network	24-FC-3	T F1-F4 L	Aircraft: It causes the loss of all DC electrical equipment and prevents the aircraft to continue safe flight or landing. Flight crew: It causes their disability to control the aircraft. Occupants: It may cause casualties of all occupants.	Catastrophic	Qualitative and quantitative FMEA, Qualitative and quantitative FTA,CCA	Note 1	
	Loss of normal DC network	24-FC-4	T F1-F4 L	Aircraft: Part of DC electrical equipment cannot work, greatly reducing the safety margin of aircraft power supply. Flight crew: It greatly increases the workload of the crew. Occupants: No effect.	Major	Qualitative and quantitative FMEA, Qualitative and quantitative FTA,CCA		Note 2
	...	...	...	...	...		...	...

Note 1 For the catastrophic failure condition, it is not required to provide supporting materials because it is the most severe failure condition, which has been designed according to the safety objectives associated with the catastrophic failure conditions without requiring a demonstration.

Note 2 These major or minor failure conditions are branches of some catastrophic and hazardous failure conditions. Quantitative analyses are required for catastrophic and hazardous failure conditions, with expansion of these failure conditions as well.

Table 3.3 Summary of FHA

FC	FC REF.	Flight phase	Classification	Verification methods
Total Loss of AC Network	24–FC–1	T F1–F4 L	Catastrophic	Qualitative and quantitative FMEA, Qualitative and quantitative FTA, CCA
Total Loss of DC Network	24–FC–3	T F1–F4 L	Catastrophic	Qualitative and quantitative FMEA, Qualitative and quantitative FTA, CCA
. . .	. . .	. . .	. . .	. . .

commence an assessment, of the failure conditions from AFHA, aircraft design requirements, initial operational considerations, proposed aircraft architecture, and so on. Once the available input data have been gathered, PASA identifies the interdependencies of the system functions that contribute to each aircraft level failure condition. The interdependence analysis can be used to validate that the functional independence and separation requirements have been adequately identified. Next, the aircraft level failure conditions are evaluated by combining the system failures that contribute to the aircraft level function by using Combined Functional Failure Effects Analysis, Multifunction & Multisystem Analysis, Common Resource Considerations Analysis, and Function Development Assurance Level (FDAL) Assignment. The evaluation of the aircraft level failure conditions helps derive the safety and design requirements for the various systems to verify that the aircraft level system architecture can reasonably be expected to meet the aircraft level safety requirements. Subsequently, PASA should confirm whether the requirements associated with each individual aircraft level failure condition can be met and that the necessary associated system requirements are derived. Finally, the PASA results are documented in a manner that provides sufficient analytical information to validate the safety requirements derived from PASA and may be used for some validation records.

EPS supplies power to all electrical equipments, which has a significant effect on the interface of other systems. The Common Cause Analysis (CCA), which includes the relevant analysis similar to PASA, will be performed in other layers of EPS. Therefore, we will not provide detailed analysis of EPS in this section.

REFERENCES

[1] SAE ARP 4761. Guidelines and methods for conducting the safety assessment process on civil airborne systems and equipment. SAE International; 1996.
[2] AC25.1309-1B (Arsenal). System design and analysis. FAA; 2002.

CHAPTER 4

System Functional Hazard Assessment

Contents

4.1 CONCEPT

As described in Chapter 3, Aircraft Functional Hazard Assessment, Functional Hazard Assessment (FHA) is a systematic and comprehensive evaluation method. It should be implemented in the early stage of the

Civil Aircraft Electrical Power System Safety Assessment
DOI: http://dx.doi.org/10.1016/B978-0-08-100721-1.00004-2

development process and updated in a timely manner when new functions or failure conditions are identified. Therefore, an FHA report is a dynamically updated document throughout the overall development process of an aircraft.

The objective of System Functional Hazard Assessment (SFHA) is to identify failure conditions, effects and classifications according to the loss of function and malfunction with the consideration of all the system functions to determine safety requirements. If the effects and classifications of failure conditions vary with the flight phases, environment, external events, and related systems, they should be identified according to different failure scenarios.

SFHA is used to allocate the safety requirements for the systems involved and to provide failure condition information, which is critical to the system safety. The information is used to determine the architecture and the requirements for the separation and isolation of the system.

SFHA is the first step of the system level safety assessment and determines the safety requirements for system development. It is an important part of the system development activities.

4.1.1 The Relationship Between System Functional Hazard Assessment and Other Safety Assessment Activities

The outputs of Aircraft Functional Hazard Assessment (AFHA) and Preliminary Aircraft Safety Assessment (PASA) are the inputs to SFHA. The function and failure condition lists of AFHA directly provide input information for SFHA. PASA provides the interface with other systems for SFHA. In the SFHA process, the traceability between functions and failure conditions of the system level and the aircraft level should be established. The classifications of the system level failure conditions should be consistent with that of the aircraft level failure conditions. If they are different, the rationale should be fed back to the aircraft level.

Note: PASA is a newly added process of SAE ARP4754A revision. The process is to examine the aircraft architecture to ensure that it meets the aircraft level safety objectives associated with the classifications of failure conditions in AFHA.

The outputs of SFHA are the inputs to Preliminary System Safety Assessment (PSSA). The analysis depth of PSSA depends on the classifications of failure conditions identified in SFHA and the relevant design and complexity of the system. For minor and no-safety-effect failure conditions, a summary in SFHA is sufficient. Catastrophic, hazardous, and some major failure conditions need further analysis in PSSA and will be analyzed in different levels of depth according to the classifications of failure conditions and the complexity of the design.

Additional attention should be paid to the validation and verification of the safety requirements and relative assumptions concluded from failure condition analysis during the SFHA process. As these activities run throughout the overall process of system development, attentions should be paid to the traceability and feedbacks between the relevant verification & validation activities and SFHA to ensure the correctness and completeness of the final SFHA results.

4.2 SYSTEM FUNCTIONAL HAZARD ASSESSMENT INPUTS

The first step of the SFHA process is to obtain the necessary data [1], which may include:
- System function list
- Functional diagram illustrating the external interfaces
- Functions and classifications of failure conditions from upper level FHA
- Requirements identified in design and objective documents
- Design decisions of the upper level

Note: Depending on whether the current level is the system level or subsystem level, the upper level could be the aircraft level or system level.

4.2.1 System Function List

The system functions are identified and listed by aircraft and system engineers and used by system engineers and safety engineers to analyze failure conditions during the SFHA process.

System functions include internal and external functions. Internal functions refer to the main functions of the system and the interface between subsystems/components; external functions, which include functions provided by the system or for the system, refer to the interface functions between the system and other aircraft systems or ground systems.

The identification of the system level functions should be based on the AFHA analysis, system design and users' requirements, learning from the engineering experience of similar aircraft, and listening to the opinions and suggestions of experts from departments of engineering, airworthiness, and airlines.

Principles for identifying the system functions are summarized as follows:

1. Function analysis can be carried out layer by layer to find all functions or subfunctions during all the operational conditions and modes.
2. Usually carried out pertaining to only the analysis object, the specific equipment or architecture for realizing functions is not involved.
3. Experts of related professions should participate in the identification of the functions.
4. The function list of a similar aircraft can be referenced when identifying the functions.

The identification of the system functions should be layered and generally divided into two layers. To ensure the completeness, correctness, and traceability of the function allocation process, system functions should correspond to the lowest level functions of the aircraft level. The lowest functions of the system level can directly correspond to specific subsystems (it is recommended that each lowest function of the system level only correspond to one subsystem).

In general, one system may perform multiple aircraft functions. This should be carefully examined and reviewed when identifying system functions to prevent omission and duplication.

The system level functions may also include some derived functions. The derived functions cannot be directly traced to the aircraft level functions. In this case, communication with the aircraft manufacturer is required to make the final determination.

When the system function is allocated to software and hardware, a new function list identified by the specific architectural design should be updated.

The traceability between system functions and aircraft functions should be established during the process of identifying the system functions.

4.2.2 Functional Diagram Showing the External Interfaces

The interface functional diagrams among systems are also an input to SFHA. The relationship among systems should be considered when analyzing the external function failures.

4.2.3 Functions and Classifications of Failure Conditions from Upper Level Functional Hazard Assessment

The functions, failure conditions, and their classifications identified in upper level FHA should be listed as input to SFHA. This input information should be referenced when identifying the failure conditions of SFHA.

4.2.4 Requirements Identified in Design and Objectives Documents

The relevant airworthiness regulations and the company's internal design requirements and objectives should be considered as input to SFHA. This information may have influences on the identification of the safety and design objectives of failure conditions in the SFHA process.

4.2.5 Design Decisions of the Upper Level

The design decisions and their descriptions of the upper level should be considered as input to SFHA because they will affect the FHA results of the analyzed system.

For example, when performing the FHA of a satellite navigation system, the safety requirements of the overall navigation system, which includes the satellite navigation system, inertial navigation system, radio navigation system, and magnetic compass system, should be taken into account. That is to say, the safety requirements allocation process of any subsystem in the navigation system is derived from the overall navigation system. Therefore, comprehensive consideration should be given to the safety design of the satellite navigation system and other subsystems to ensure that the overall navigation system meets the safety requirements of the aircraft.

4.3 SYSTEM FUNCTIONAL HAZARD ASSESSMENT PROCESS

SFHA is a top-down logical approach. The overall process of SFHA includes:

1. Identification and description of failure conditions.
2. Determination of the effects and classifications of failure conditions.
3. Allocation of safety objectives to failure conditions.
4. Identification of verification methods for the compliance with the requirements of failure conditions.
5. Identification of supporting materials for the classifications of failure conditions.

The following sections of this chapter will analyze and illustrate the above-mentioned points.

4.3.1 Identification and Description of Failure Conditions

4.3.1.1 Environmental Conditions and Emergency Configuration

Failure conditions are closely related to environmental conditions and emergency configuration. It is necessary to consider the effects on failure conditions caused by environmental conditions and emergency configurations.

Considering the environmental conditions makes the effects of failure conditions more severe, for example, when comparing the two failure conditions of "Loss of the Anti-icing Function" and "Loss of the Anti-icing Function Under the Icing Condition", the effects are different based on consideration (or not) of the icing condition. Obviously, considering the icing condition will make the classification of the failure condition into a higher level. For another example, the failure condition "Loss of the Detection Function" of the engine fire detection system has a minor effect, while this effect will be catastrophic in the case of considering the engine fire.

Examples of environmental conditions considered in SFHA include the following:

- Icing conditions
- Lightning and High Intensity Radiated Fields (HIRF)
- Polluted runway
- Fire
- Bird strike, etc.

It is also necessary to list the emergency configurations resulted from emergency/abnormal conditions that need to be considered when determining the failure condition effects. There are two types of emergency configurations. One is derived from the emergency configurations of AFHA, which includes the following:

- Ditching
- Emergency landing
- Engine out
- Loss of communications
- Depressurization

In fact, the configuration derived from AFHA usually refers to the configuration of a latent-related system. Here, "latent" specifically means that there is no direct interface, but once this configuration occurs, it will have an effect on SFHA of the considered system.

The other is derived from the system architecture design in the initial design stage. These emergency configurations often have direct interfaces with the considered system, such as:

- Loss of the Hydraulic System
- Loss of the Electrical System
- Loss of the Equipment Cooling System

It should be noted that not all the systems are related to environmental conditions and emergency configurations.

4.3.1.2 Identification of Failure Conditions Due to Single or Combined Function Failures

It is necessary to identify failure conditions due to single or combined function failures for each function in the system function list.

4.3.1.2.1 Identification of Failure Conditions Due to Single Function Failure

Typical single function failure include the following:

1. Loss of Function: loss of the expected function

    Loss of function can be further divided into the following:
    - Complete loss of function: complete loss of the ability to conduct the expected system functions
    - Partial loss of function: partial loss of the ability to conduct the expected system functions

Note: Some functions may not include partial loss. In this case, only the complete loss of function should be considered.

2. Malfunction: the function is outside the limits of its normal operation
   The following are examples of malfunction:
   - Actions without commands: conducting the expected or unexpected system function without commands
   - Out of limits: some condition exceeds the expected upper or lower limit

4.3.1.2.2 Identification of Failure Conditions Due to Combined Function Failures

The effects of combined function failures may usually be more severe than any single function failure therein. Therefore, in addition to the consideration of single function failures, different combined function failures should also be assumed when identifying failure conditions. The purpose of considering combined function failures is to identify whether there are more severe failure conditions; if the severity of combined function failures is the same as any single function failure therein, the consideration of combined conditions is no longer necessary.

Combined function failures are more complex and require analysts to have a comprehensive understanding of system architecture, main components, and the relationship between internal and external system functions. Combined function failures should be considered in the following aspects:

- Associated with failures of other functions (e.g., warnings, protections, etc.)
- Associated with particular flight phases (e.g., takeoff, landing, etc.)
- Associated with environmental conditions and emergency configurations (e.g., icing, fire, depressurization, etc.)
  Typical examples of combined function failures are as follows:
- Loss of communication and navigation
- Loss of two hydraulic systems

To maintain good traceability, failure conditions could be marked. It is suggested that failure conditions be marked by way of "function number + subnumber." For example, a function of the Electrical Power System (EPS) is referenced as "24-1," so the failure condition corresponding to this function could be "24-1-1," and then the following failure conditions are numbered sequentially.

4.3.2 Determination of the Effects and Classifications of Failure Conditions

4.3.2.1 The Effects of Failure Conditions in Different Flight Phases

For the same failure condition, the failure effects and classifications may vary from one flight phase to another. FHA should identify the effects of failure conditions for each flight phase.

The division of flight phases in SFHA should be consistent with AFHA. Flight phases should be divided carefully according to engineering experience and aircraft design requirements because it will affect the safety objectives in SFHA. For a system supplier, the division of flight phases should be fully consistent with the aircraft manufacturer.

For example, the division of flight phases of an aircraft is usually as follows:

A. Ground
  G1: Taxiing
  G2: Power on on ground
B. Takeoff
  T1: Takeoff
  T2: Rejected takeoff
C. In-flight
  F1: Climb
  F2: Cruise
  F3: Approach
  F4: Go around
D. Landing
  L1: Landing roll

4.3.2.2 Determination of the Effects of Failure Conditions

For each failure condition, functional failure analysis should be conducted to determine *the worst effects* (regardless of their probability) of failure conditions on the aircraft, flight crew, and occupants. It is important to determine the effects of failure conditions, which are the basis for the classifications of failure conditions and the allocation of safety requirements.

Detailed analysis of the effect of each failure condition on the aircraft, flight crew, and occupants could include the methods of failure detection, the corrective actions of the crew, and the aircraft situation after actions taken by the crew. It is necessary to determine the effects of failure conditions for the following three aspects:

4.3.2.2.1 Effects on the Aircraft

Determine the effects of failure conditions on the aircraft and system. Not only all the direct and immediate effects but also the subsequent effects are taken into consideration herein. If the effects vary from one flight phase to another, all cases should be considered.

Note: "Subsequent effects" are not the direct consequence of failure conditions but instead induced by the correction actions of the crew and other factors, such as flight environment, Air Traffic Control (ATC), etc.

The description of the effects of failure conditions can be divided into the following two aspects:
1. To describe what the effects on the aircraft are;
2. To summarize the severity of their effects in the following five categories.
    * Loss of the aircraft
    * A large reduction in safety margins or functional capabilities
    * A significant reduction in safety margins or functional capabilities
    * A slight reduction in safety margins or functional capabilities
    * No effect on safety

4.3.2.2.2 Effects on the Flight Crew

1. To describe the failure detection methods and determine whether it is a hidden failure and what corrective actions should be taken by the flight crew.

    Failure Detection Methods: All indications should be described to the flight crew and the ground staff so that the failure will be detectable, and much easier so. The indications include Engine Indication and

Crew Alerting System (EICAS), instruments, visual information, voice, and the feeling of the human body.

Hidden failures mean the failures that are not detected and/or annunciated when it occurs and should also be determined here.

Corrective Action Taken by the Flight Crew: The expected corrective actions taken by the flight crew could be specifically illustrated and the appropriate procedures could be carried out when there are warning indications. If it is applicable, the specific training content for flight crews could be determined or referenced (if any).

2. To summarize the effects via the following five categories.
   * Fatalities or incapacitation
   * Physical distress or excessive workload such that the flight crew cannot be relied upon to perform their tasks accurately or completely
   * Significant increase in flight crew workload or in conditions that impair crew efficiency or cause discomfort to the flight crew
   * A slight increase in flight crew workload, such as routine flight plan changes
   * No effect

4.3.2.2.3 Effects on the Occupants

According to AC25.1309-1B (Arsenal), the effects on the occupants can be classified into the following five categories:
* Multiple fatalities
* Serious or fatal injury to a relatively small number of occupants
* Physical distress to the occupants, possibly including injuries
* Some physical discomfort to the occupants
* No effect

4.3.2.3 Determination of the Classification of Failure Condition Effects

According to AC25.1309-1B (Arsenal) [3], failure conditions can be divided into five classifications, namely, Catastrophic, Hazardous, Major, Minor, and No Safety Effect. Their detailed definitions are given in Chapter 1, Airworthiness Regulations and Safety Requirements.

The classifications of failure conditions does not depend on whether the systems and components are required by the relevant regulations (e.g., Part 25, etc.); it only depends on the severity of the system failure or functional failure. For example, the installation of the position light system is required by Part 25, but the severity of failure conditions is generally minor. For those systems that are not required by Part 25, such as the flight management system and the automatic landing system, the severity of failure conditions might be major or catastrophic.

Based on the effects of failure conditions and the definition and description of the five categories above, the classification of failure condition effects is determined.

The classification of failure condition effects can depend on the analysis of accidents/incidents data, guidance materials, and previous design experience, and the flight crew can also be consulted. For their respective working roles, different people have different judgment regarding the classifications of failure conditions. Take the Traffic Collision Avoidance System (TCAS II) as an example: the flight crew may consider the failure of the TCAS function in hub airports as likely to lead to collision, while the system designers may think that the effect can be minor since the aircraft still can be controlled even without TCAS function, and of course, easier for design implementation. However, the failure condition of the TCAS function in TSO-C119d [2] is hazardous.

4.3.2.4 Combination of Failure Conditions

The combination of failure conditions can be divided into two situations as follows:

1. Different failure conditions but the same effects

   Different failure conditions can lead to the same effects on the aircraft and system. They can be grouped together under the same failure condition (with a unique reference and title).
2. The same failure conditions, but the effects vary for different flight phases

   For the same failure conditions but varying effects based on the flight phase, there are two ways to deal with it:
   a. Consider them as different failure conditions
   b. Combine them to a new failure condition sharing a common title, list the most severe effects as the effects of the new failure condition, and list all relevant flight phases.

The second method increases the exposure time, and therefore the safety requirements are raised artificially and the result is somewhat conservative. It is recommended that the second method can be considered preferentially because it will reduce the number of failure conditions and the later workload. When the second method cannot meet the safety requirements, the first method is then considered.

4.3.3 Allocation of Safety Objectives to Failure Conditions

The analyst should allocate quantitative safety objectives and qualitative design requirements to each failure condition based on its effects and classification. The quantitative safety objective is the maximum allowable probability for the failure condition. It is possible to set a design objective more stringent than safety objectives to take into account reliability, economic repercussions, state of the art, etc. The approval of this design objective is not part of the certification process.

4.3.3.1 Safety Objectives

Safety objectives include quantitative and qualitative objectives.

Quantitative Objective: The maximum allowable probability of failure conditions. The quantitative objectives of different types of the aircraft are different. For example, the transport aircraft should refer to AC25.1309-1B (Arsenal), and the relationship between failure conditions and safety objectives is shown in Table 4.1. For normal, utility, acrobatic, and commuter category airplanes, refer to AC23.1309-1E, *Safety Analysis and Assessment for Part System 23 Airplanes.*

Qualitative objectives: Fail-safe, Development Assurance Level (DAL), architecture mitigation measures, etc.

4.3.3.2 Design Objectives

Table 4.2 shows an example of the effects of classifications and objectives of failure conditions on the operational reliability, which is based on experience from aircraft manufacturers and airlines.

4.3.4 Identification of Verification Methods for the Compliance with the Requirements of Failure Conditions

For each failure condition, the analyst should determine how the aircraft/system will meet the safety objective, so the verification methods should be identified. There are two aspects. First, determine whether qualitative analysis, quantitative analysis, or the combination of the two should be used [4]. Second, identify which safety assessment method should be used. For example, verification methods of a catastrophic failure condition are qualitative Failure Modes and Effects Analysis (FMEA), qualitative Fault Tree Analysis (FTA), quantitative FMEA, quantitative FTA, and Common Cause Analysis (CCA).

Table 4.1 Relationship between the probability and severity of failure conditions [3]

Effect on the aircraft	No effect on operational capabilities or safety	Slight reduction in functional capabilities or safety margins	Significant reduction in functional capabilities or safety margins	Large reduction in functional capabilities or safety margins	Normally with hull loss
Effect on occupants excluding the flight crew	Inconvenience	Physical discomfort	Physical distress, possibly including injuries	Serious or fatal injury to a small number of occupants	Multiple fatalities
Effect on the flight crew	No effect on the flight crew	Slight increase in workload	Physical discomfort or a significant increase in workload	Physical distress or excessive workload that impairs the ability to perform tasks	Fatalities or incapacitation
Allowable qualitative probability	No probability requirement	<...Probable...>	<...Remote...>	Extremely <..............> Remote	Extremely improbable
Allowable quantitative probability: average probability per flight hour on the order of:	No probability requirement	<..............> < 10^{-3} Note 1	<..............> < 10^{-5}	<..............> < 10^{-7}	< 10^{-9}
Classification of failure conditions	No Safety Effect	<...Minor...>	<...Major...>	<...Hazardous...>	Catastrophic

Note 1: A numerical probability range is provided here as a reference. The applicant is not required to perform a quantitative analysis, nor substantiate such by an analysis, that this numerical criterion has been met for minor failure conditions. Current transport category airplane products are regarded as meeting this standard simply by using current commonly accepted industry practices.

Table 4.2 Category and objective examples of operational reliability

Category	Description	Design objective of failure conditions (per FH)	Design objective of aircraft level (per FH)
Flight interrupt	The failure condition of starting the emergence process	$<2.0 \times 10^{-7}$	$<2.0 \times 10^{-5}$
Delay	The airplane was delayed more than 15 minutes	$<2.0 \times 10^{-5}$	$<2.0 \times 10^{-3}$

4.3.5 Identification of Supporting Materials for the Classifications of Failure Conditions

The necessary supporting material should be listed for some effect-unclear and controversial failure conditions to support the judgment of failure condition effects, classifications, flight crew actions, verification methods, etc. SFHA is implemented in the early stage of the system development process, and the material is given only with the index in this stage. Supporting materials include the following:

- Analysis
- Calculation
- Design description
- Simulator test, flight test, ground test, etc.
- Crew procedures
- Engineering experience

Note: Considering the safety of test crew, generally we do not use flight tests, but simulation methods, to justify the effects of hazardous failure conditions. However, it is possible to use flight tests to justify the effect of major.

The following two cases do not need to provide supporting materials of failure condition classifications:

- The effects of failure conditions are clear, unambiguous and agreed by common experience.
- The catastrophic failure conditions are the most severe, and the highest safety objectives have already been set, so supports are no longer necessary.

4.4 SYSTEM FUNCTIONAL HAZARD ASSESSMENT OUTPUTS

The main content of the SFHA document is the failure condition list with safety objectives. The failure condition summary is as presented in Table 4.3, or cut on the basis of Table 4.3, but it should at least include all failure conditions that relate to the safety objectives of the system.

This section should be used as a summary of the SFHA process. The critical failure conditions, including the critical operations of the aircraft and major, hazardous, or catastrophic failure conditions, are listed in Table 4.3, and they need to be further analyzed in PSSA and SSA; in this section, the safety objectives and design objectives of each failure condition are listed as well.

To ensure the traceability of SFHA, materials related to the SFHA process should be documented, such as function lists, environment and emergency configuration lists, and supporting materials.

4.5 THE CASE STUDY OF ELECTRICAL POWER SYSTEM FUNCTIONAL HAZARD ASSESSMENT

The SFHA process is illustrated below by taking the EPS as an example. This case is simplified, and due to the differences in specific projects, the analysis and the results of this SFHA case are not necessarily applicable to other SFHAs. Therefore, this case is for reference only.

Table 4.3 Failure condition summary

FC REF (1)	FC title (2)	Classification (3)	Safety objective (4)	Design objective (5)	Verification methods (6)	Cross REF with supporting materials (7)

(1) List the failure condition ID.
(2) List the title associated with the failure condition.
(3) List the classification of the effect on the aircraft in terms of safety and operational reliability.
(4) List the safety objective, which should be consistent with the classification of the failure condition.
(5) List the design objective associated with the failure condition (if it is different from the safety objective).
(6) List verification methods for requirements.
(7) List the cross-reference with supporting materials.
Note: All numerical probabilities (objectives) should normally be quoted per hour of flight.

4.5.1 Electrical Power System Description

Aircraft Description: A two engine transport aircraft designed to carry 75 passengers with an average cruising speed of 0.78 Mach, a maximum range of 1200 nautical miles, an average flight duration of 1.2 hours, and a 50,000-hour service life.

The aircraft is required to provide EPS with AC and DC power for all the electrical equipment and emergency power for essential loads when the aircraft loses power. It also needs to provide necessary power when the aircraft is on the ground with the engine shut down. Therefore, the aircraft EPS is composed of AC power, DC power, emergency power, auxiliary power, corresponding distribution, and other subsystems.

The AC power system, with a 115/200 V, 400 Hz, 3-phase power supply, is composed of the main AC power system, auxiliary AC power system, external AC power system, and emergency AC power system.

The main AC power system consists of the combination of an Integrated Drive Generator (IDG), Generator Control Unit (GCU), and Power Distribution Assembly (PDA).

The auxiliary AC power system consists of a generator driven by an APU, GCU, and PDA.

The external AC power system, with the 115/200 V, 400 Hz, 3-phase power supply, is composed of a Bus Power Control Unit, external power socket, etc.

The emergency AC power system is composed of an air-cooled generator driven by a Ram Air Turbine (RAT), RAT GCU, and static inverter (INV).

The DC power system, rated at 250 A, is composed of the secondary DC power and battery system. The secondary DC power provides 28 V of DC power, which is divided into left and right DC power systems and an emergency DC power system, composed of a Transformer Rectifier Unit (TRU) and TRU contactors. The battery system provides 24 V of DC power, which is divided into the main battery system and APU battery system.

The distribution system is composed of an AC power distribution system and a DC power distribution system. The AC power distribution system is mainly composed of left and right AC buses, the AC essential bus, and the AC ground service bus. The DC distribution system consists

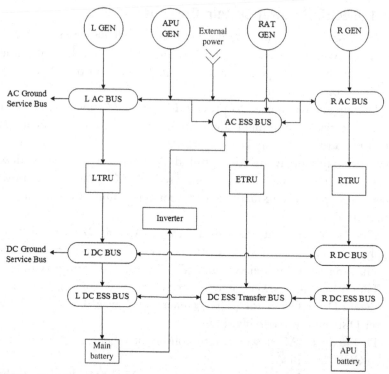

Figure 4.1 The functional block diagram of the EPS. EPS, electrical power system.

of left and right DC buses, left and right DC essential buses, the DC essential transfer bus, the DC ground service bus, etc.

The functional block diagram of the EPS is shown in Fig. 4.1.

4.5.2 Input Information

4.5.2.1 Function List of the Electrical Power System

1. The main functions of the generation system are:
   a. To provide electrical power required by the aircraft load
      i To provide AC power by the main AC generation system in normal conditions
      ii To provide DC power by the main DC generation system in normal conditions
      iii To provide emergency power by the emergency AC and DC generation system in the case of emergency
   b. To control the electrical power quality provided for electrical equipment and maintaining quality within design standards

     c. To deliver the power from the generator to the distribution system

     d. To provide system status and warning information (e.g., voltage, current, etc.)

     e. To provide system maintenance data

2. The main functions of the distribution system are:

     a. To provide buses for electrical equipment

       i To provide AC bus

       ii To provide DC bus

     b. The function of wire protection

     c. To provide management functions to control and monitor the distribution network

     d. The function of load management

     e. To provide system status and warning information

     f. To provide system maintenance data

4.5.2.2 Functions and Failure Conditions Identified by Aircraft Functional Hazard Assessment

Functions and failure conditions identified by AFHA are shown in Table 4.4.

4.5.3 Function Hazard Assessment

For the four functions, namely, "Provide AC Power for AC Electrical Equipment, 24-1", "Provide Buses between AC Power and AC Electrical Equipment, 24-2", "Provide DC Power for DC Electrical Equipment,

Table 4.4 Functions and failure conditions identified by AFHA

Function	FC	Effect of the failure condition on aircraft or personnel	Flight phase	Classification
Provide Electrical Power	Total Loss of Electrical Power	Aircraft: Loss of the aircraft. Flight crew: Flight crew is unable to control the aircraft. Occupants: May cause the death of a vast majority of occupants.	T F1–F4 L	Catastrophic

24-3", and "Provide Buses between DC Power and DC Electrical Equipment, 24-4", this case will perform FHA on them.

4.5.3.1 Failure Conditions

For the function of "Provide AC Power for AC Electrical Equipment, 24-1", the failure conditions are identified from two aspects: loss of its normal function (single function failure) and combined loss with AC backup power and emergency power (combined function failures).

> Function: Provide AC Power for AC Electrical Equipment
> Failure Conditions:
>   Loss of AC Normal, Backup and Emergency Power (combined function failures)
>   Loss of AC Normal and Backup Power (combined function failures)
>   Loss of AC Normal Power (single function failure)

For the function "Provide Bus between AC Power and AC Electrical Equipment, 24-2", the failure conditions are identified from three aspects: loss of all AC buses (combined function failures), loss of multiple AC buses (combined function failures), and loss of one AC bus (single function failure).

> Function: Provide Buses between AC Power and AC Electrical Equipment
> Failure Conditions:
>   Total Loss of AC Buses (combined function failures)
>   Loss of Multiple AC Buses (combined function failures)
>   Loss of One AC Bus (single function failure)

For the function "Provide DC Power for DC Electrical Equipment, 24-3", the failure conditions are identified from two aspects: loss of its normal AC bus (single function failure) and combined loss with DC emergency (combined function failures).

> Function: Provide DC Power for DC Electrical Equipment
> Failure Conditions:
>   Loss of DC Normal and Emergency Power (combined function failures)
>   Loss of DC Normal Power (single function failure)

For the function "Provide Buses between DC Power and DC Electrical Equipment, 24-4", the failure conditions are identified from three aspects: loss of all DC buses (combined function failures), loss of multiple DC buses (combined function failures), and loss of one DC bus (single function failure).

Function: Provide Buses between DC Power and DC Electrical Equipment
Failure Conditions:
    Total loss of DC Buses (combined function failures)
    Loss of Multiple DC Buses (combined function failures)
    Loss of One DC Bus (single function failure)

4.5.3.2 Flight Phase
A. Ground
    G1: Taxiing
B. Takeoff
    T1 Takeoff run (before V1)
    T2 Takeoff
    T3 Rejected takeoff
C. In-flight
    F1: Climb
    F2: Cruise
    F3: Decline
    F4: Approach
D. Landing
    L: Landing roll
E. ALL: refers to all of the above phases.
    The failure conditions are shown in Table 4.5.

4.5.3.3 Determine the Effects of Failure Conditions on the Aircraft, Flight Crew, and Occupants
For each failure condition, its effects on the aircraft, flight crew, and occupants are determined as in Section 4.3.2.2. See Table 4.5 for details.

4.5.3.4 Determine the Classifications of Failure Conditions on the Aircraft, Flight Crew, and Occupants
For each failure condition, the classification of its effects on the aircraft, flight crew, and occupants is determined as in Section 4.3.2.3. See Table 4.5 for details.

Table 4.5 Worksheet example for the EPS in SFHA

Function	FC	FC REF.	Flight phase	Effect of the FC on aircraft or personnel 1. Aircraft 2. Flight crew 3. Occupants	Classification	Verification methods	Reference to supporting material	Remarks
Provide AC Power for AC Electrical Equipment (24-1)	Loss of AC Normal, Backup and Emergency Power (L AC Generator and R AC Generator, APU Generator, RAT Generator)	24-1-1	T F1–F4 L	Aircraft: It causes the loss of functionality of all AC electrical equipment and prevents the aircraft from continuing safe operation or landing. Flight crew: It causes their inability to control the aircraft. Occupants: It may cause casualties of all occupants.	Catastrophic	Qualitative and quantitative FMEA, Qualitative and quantitative FTA, CCA	Note 1	
	Loss of AC Normal and Backup Power (L AC Generator and R AC Generator, APU Generator)	24-1-2	T F1–F4 L	Aircraft: It causes the loss of functionality because related power cannot be supplied to the two electrical units, the AC normal and backup power units (e.g., the primary flight control system and hydraulic system required for AC power, as well as the TRU providing power for associated DC electrical equipment). However, this electrical equipment can be supplied with emergency power. This failure will extremely reduce the safety margin of the aircraft power supply. Flight crew: It extremely increases the workload of the crew. Occupants: In case of the functional loss of the pressurization, air-conditioner, and oxygen systems, it may cause discomfort and even threaten the lives of some weaker occupants.	Hazardous	Qualitative and quantitative FMEA, Qualitative and quantitative FTA, CCA		

(Continued)

Table 4.5 (Continued)

Function	FC	FC REF.	Flight phase	Effect of the FC on aircraft or personnel 1. Aircraft 2. Flight crew 3. Occupants	Classification	Verification methods	Reference to supporting material	Remarks
	Loss of AC Normal Power (L AC Generator and R AC Generator)	24-1-3	T F1–F4 L	Aircraft: It causes the loss of functionality because the electrical units (e.g., the primary flight control system and the provision of power for associated DC electrical equipment) supplied with AC normal power cannot receive power from it. However, this electrical equipment can be supplied with backup power. This failure will greatly reduce the safety margin of the aircraft power supply. Flight crew: It greatly increases the workload of the crew. Occupants: No effect.	Major	Qualitative and quantitative FMEA, Qualitative and quantitative FTA, CCA		Note 2
Provide AC Buses Between AC Power and AC Electrical Equipment (24-2)	Total Loss of AC Buses							
	Loss of the AC ESS Bus, L AC Bus, and R AC Bus	24-2-1	T F1–F4 L	Aircraft: None of the AC electrical equipment can work, preventing the aircraft from continuing safe operation or landing. Flight crew: It causes their inability to control the aircraft. Occupants: It may cause casualties of all occupants.	Catastrophic	Qualitative and quantitative FMEA, Qualitative and Quantitative FTA, CCA	Note 1	
	Loss of Multiple AC Buses							
	Loss of the L AC Bus and R AC Bus	24-2-2a	T F1–F4 L	Aircraft: The AC electrical equipment connected to the bus cannot work while the AC ESS electrical equipment can only get power from the AC ESS bus, greatly reducing the safety margin of the aircraft power supply. Flight crew: It greatly increases the workload of the crew.	Major	Qualitative and quantitative FMEA, Qualitative and quantitative FTA, CCA		Note 2

(Continued)

Table 4.5 (Continued)

Function	FC	FC REF.	Flight phase	Effect of the FC on aircraft or personnel 1. Aircraft 2. Flight crew 3. Occupants	Classification	Verification methods	Reference to supporting material	Remarks
	Loss of the L AC Bus and AC ESS Bus	24-2-2b	T F1-F4 L	Occupants: No effect. Aircraft: The AC electrical equipment connected to the bus cannot work, greatly reducing the safety margin of the aircraft power supply. Flight crew: It extremely increases the workload of the crew. Occupants: In case of the functional loss of pressurization, air-conditioner and oxygen systems, it may cause discomfort and even threaten the lives of some weaker occupants.	Hazardous	Qualitative and Quantitative FMEA, Qualitative and Quantitative FTA, CCA		
	...	...	...	...	...	...	...	...
Loss of One AC Bus								
	Loss of the L AC Bus	24-2-3a	T F1-F4 L	Aircraft: The AC electrical equipment connected to the L AC bus cannot work, greatly reducing the safety margin of the aircraft power supply. Flight crew: It greatly increases the workload of the crew. Occupants: No effect.	Major	Qualitative and quantitative FMEA, Qualitative and quantitative FTA		Note 2
	Loss of the R AC Bus	24-2-3b	T F1-F4 L	Aircraft: The AC electrical equipment connected to the R AC bus cannot work, greatly reducing the safety margin of the aircraft power supply. Flight crew: It greatly increases the workload of the crew. Occupants: No effect.	Major	Qualitative and quantitative FMEA, Qualitative and quantitative FTA		Note 2

(Continued)

Table 4.5 (Continued)

Function	FC	FC REF.	Flight phase	Effect of the FC on aircraft or personnel 1. Aircraft 2. Flight crew 3. Occupants	Classification	Verification methods	Reference to supporting material	Remarks
	Loss of the AC ESS Bus	24-2-3c	T F1–F4 L	Aircraft: The AC electrical equipment connected to the AC ESS bus cannot work, greatly reducing the safety margin of the aircraft power supply. Moreover, it may cause the functionality loss of some important electrical equipment (for nonredundant systems). Flight crew: It greatly increases the workload of the crew. Occupants: No effect.	Major	Qualitative and quantitative FMEA, Qualitative and quantitative FTA		Note 2
	…	…	…	…	…	…	…	…
Provide DC Power for DC Electrical Equipment (24-3)	Loss of DC Normal and Emergency Power (LTRU, RTRU, ETRU)	24-3-1	T F1–F4 L	Aircraft: For DC electrical equipment connected to DC normal and emergency power, its power is supplied by the main battery and APU battery for a limited time. It is unable to ensure that the aircraft can continue long-term safe operation or land if the time limit is exceeded. Flight crew: It causes their inability to control the aircraft if the time limit is exceeded. Occupants: It may cause casualties of all the occupants.	Catastrophic	Qualitative and quantitative FMEA, Qualitative and quantitative FTA, CCA	Note 1	

(Continued)

Table 4.5 (Continued)

Function	FC	FC REF.	Flight phase	Effect of the FC on aircraft or personnel 1. Aircraft 2. Flight crew 3. Occupants	Classification	Verification methods	Reference to supporting material	Remarks
	Loss of DC Normal Power (LTRU, RTRU)	24-3-2	T F1-F4 L	Aircraft: The DC electrical equipment (e.g., the primary flight control system, engine electronic control system, communication and navigation system, air system, environmental system, indication and recording system, warning system, and integrative processing system) connected to the DC normal power cannot work, greatly reducing the safety margin of the aircraft power supply. In the case of main battery failure and APU battery failure, the short-term (transient) power supply is lost during the conversion of DC emergency power to the L DC ESS bus and R DC ESS bus, thus causing the malfunction of the DC electrical equipment of the indirect L DC ESS bus and R DC ESS bus. Flight crew: It greatly increases the workload of the crew. Occupants: No effect.	Major	Qualitative and quantitative FMEA, Qualitative and quantitative FTA, CCA		Note 2
	…	…	…	…	…	…	…	…
Provide Buses Between DC Power and DC Electrical Equipment (24-4)	Total Loss of DC Buses							
	Loss of the L DC Bus, R DC Bus, DC ESS Bus, and DC ESS Transfer Bus	24-4-1	T F1-F4 L	Aircraft: None of the DC electrical equipment can work, preventing the aircraft from continuing safe operation or landing. Flight crew: It causes their inability to control the aircraft. Occupants: It may cause casualties of all the occupants.	Catastrophic	Qualitative and quantitative FMEA, Qualitative and quantitative FTA, CCA	Note 1	

(Continued)

Table 4.5 (Continued)

Function	FC	FC REF.	Flight phase	Effect of the FC on aircraft or personnel 1. Aircraft 2. Flight crew 3. Occupants	Classification	Verification methods	Reference to supporting material	Remarks
Loss of Multiple DC Buses								
	Loss of the L DC Bus, R DC Bus, and DC ESS Transfer Bus	24-4-2a	T F1-F4 L	Aircraft: The DC electrical equipment connected to these buses cannot work, preventing the aircraft from continuing safe operation or landing. Flight crew: It causes their inability to control the aircraft. Occupants: It may cause casualties of all the occupants.	Catastrophic	Qualitative and quantitative FMEA, Qualitative and quantitative FTA, CCA	Note 1	
	Loss of the L DC Bus and R DC Bus	24-4-2b	T F1-F4 L	Aircraft: The DC electrical equipment connected to these buses cannot work, greatly reducing the operational ability or safety margin of the aircraft. Under this condition, important redundant systems should be at least connected to the L or R DC bus and L or R DC ESS bus to ensure the functionality of the DC ESS electrical equipment. Flight Crew: It greatly increases the workload of the crew. Occupants: No effect.	Major	Qualitative and quantitative FMEA, Qualitative and quantitative FTA, CCA		Note 2
	Loss of the L DC ESS Bus and R DC ESS Bus	24-4-2c	T F1-F4 L	Aircraft: The DC electrical equipment connected to these buses cannot work, greatly reducing the operational ability or safety margin of the aircraft (e.g., if a catastrophic electrical system is a thrice-redundant system, at least one channel should be connected to the L or R bus, and another two channels should be connected to the L and R DC ESS bus to ensure the functionality of the DC ESS electrical equipment). Flight crew: It extremely increases the workload of the crew.	Hazardous	Qualitative and quantitative FMEA, Qualitative and quantitative FTA, CCA		Note 3

(Continued)

Table 4.5 (Continued)

Function	FC	FC REF.	Flight phase	Effect of the FC on aircraft or personnel 1. Aircraft 2. Flight crew 3. Occupants	Classification	Verification methods	Reference to supporting material	Remarks
				Occupants: In the case of the functional loss of pressurization, air-conditioner and oxygen systems, it may cause discomfort and even threaten the lives of some weaker occupants.				
	...	...	...	...	...	...	...	...
Loss of One DC Bus								
Loss of the L DC Bus	24-4-3a	T F1–F4 L	Aircraft: The DC electrical equipment connected to the L DC bus cannot work, greatly reducing the operational ability or safety margin. Flight crew: It greatly increases the workload of the crew. Occupants: No effect.	Major	Qualitative and quantitative FMEA, Qualitative and quantitative FTA		Note 2
Loss of the R DC Bus	24-4-3b	T F1–F4 L	Aircraft: The DC electrical equipment connected to the R DC bus cannot work, greatly reducing the operational ability or safety margin. Flight crew: It greatly increases the workload of the crew. Occupants: No effect.	Major	Qualitative and quantitative FMEA, Qualitative and quantitative FTA		Note 2
...	...	...	...	...	...	...	...	...

Note 1: For the failure condition of catastrophic, it is not required to provide supporting materials because it is the most severe failure condition, which has been designed according to the safety objective associated with the catastrophic failure conditions without requiring a demonstration.

Note 2: These major or minor failure conditions are branches of some catastrophic and hazardous failure conditions. Quantitative analyses are required for catastrophic and hazardous failure conditions, with expansion of these failure conditions as well.

Note 3: If the electrical system has two redundant channels which are respectively connected to the L and R DC ESS buses, the classification of this failure condition should be catastrophic.

4.5.3.5 Determine the Verification Methods for Demonstrating That the Safety Objective Has Been Met

For each failure condition, the verification methods for demonstrating that the safety objective has been met is determined as in Section 4.3.4. See Table 4.5 for details.

4.5.3.6 The Combination of Failure Conditions

In Table 4.5, the failure condition "Loss of Normal, Backup and Emergency Power, 24-1-1" and the failure condition "Loss of the AC ESS Bus, L AC Bus, and R AC Bus, 24-2-1" both lead to the functional loss of all AC electrical equipment and prevent the aircraft from continuing safe operation or landing, and their effects are catastrophic. Therefore, the two failure conditions can be combined into a new failure condition entitled "Total Loss of AC Network" with a new ID, which is also a catastrophic failure condition.

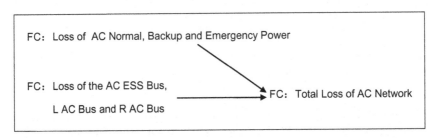

The failure condition "Loss of AC Normal Power, 24-1-3" and the failure condition "Loss of the L AC Bus and R AC Bus, 24-2-2a" both cause that part of the AC electrical equipment to stop working, which will greatly reduce the safety margin of the aircraft power supply, and their effects are major. Therefore, the two failure conditions can be combined into a new failure condition titled "Loss of AC Normal Network" with a new ID, which is also a major failure condition.

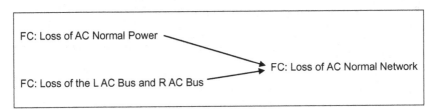

The failure condition "Loss of DC Normal and Emergency Power, 24-3-1", the failure condition "Loss of the L DC Bus, R DC Bus, DC ESS Bus, and DC ESS Transfer Bus, 24-4-1", and the failure condition "Loss of the L DC Bus, R DC Bus, and DC ESS Transfer Bus, 24-4-2a" all cause all of the DC electrical equipment to stop working, preventing the aircraft from continuing safe operation or landing, and their classifications are catastrophic. Therefore, the three failure conditions can be combined into a new failure condition entitled "Total Loss of DC Network" with a new ID, which is also a catastrophic failure condition.

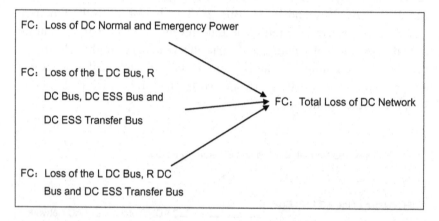

Similarly, the failure condition "Loss of DC Normal Power, 24-3-2" and the failure condition "Loss of the L DC Bus and R DC Bus, 24-4-2b," both cause that part of the DC electrical equipment to stop working, greatly reducing the operational ability or safety margin of the aircraft, and thus their effects are major. Therefore, the two failure conditions can be combined into a new failure condition entitled "Loss of DC Normal Network" with a new ID, which is also a major failure condition.

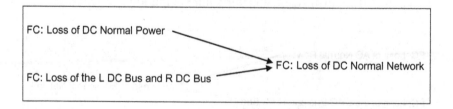

Table 4.6 EPS FHA summary table

FC	FC REF.	Flight phase	Classification	Verification methods
Total Loss of AC Network	24–FC-1	T F1–F4 L	Catastrophic	Qualitative and quantitative FMEA, Qualitative and quantitative FTA, CCA
Loss of AC Normal Network	24–FC-2	T F1–F4 L	Major	Qualitative and quantitative FMEA, Qualitative and quantitative FTA, CCA
Total Loss of DC Network	24–FC-3	T F1–F4 L	Catastrophic	Qualitative and quantitative FMEA, Qualitative and quantitative FTA, CCA
Loss of DC Normal Network	24–FC-4	T F1–F4 L	Major	Qualitative and quantitative FMEA, Qualitative and quantitative FTA, CCA
...	...	...	...	...

4.5.4 The Summary of System Functional Hazard Assessment

According to the results of SFHA analyses, the summary of FHA on the EPS is shown in Table 4.6.

REFERENCES

[1] SAE ARP 4761. Guidelines and methods for conducting the safety assessment process on civil airborne systems and equipment. SAE International; 1996.
[2] TSO-C119d. Traffic alert and collision avoidance system (TCAS) airborne equipment, TCAS II with hybrid. FAA; 2013.
[3] AC 25.1309-1B (Arsenal). System design and analysis. FAA; 2002.
[4] AC 25.1309-1A. System design and analysis. FAA; 1988.

CHAPTER 5

Preliminary System Safety Assessment

Contents

Civil Aircraft Electrical Power System Safety Assessment
DOI: http://dx.doi.org/10.1016/B978-0-08-100721-1.00005-4
101

5.1 INTRODUCTION

Preliminary System Safety Assessment (PSSA) is used to assess the system architecture, determining how failures lead to failure conditions in Functional Hazard Assessment (FHA) and how to meet safety objectives. It will allocate the system level safety requirements (failure probability, Development Assurance Level (DAL), etc.) produced by System Functional Hazard Assessment (SFHA) to the subsystems/items and provide the necessary input for other system development activities, as shown in Fig. 5.1. PSSA is closely related to and interacts with design activities, and they are iterated in turns in the overall design cycle.

The functions and purposes of PSSA mainly include the following aspects:

1. To explore the mechanism for the occurrence of failure conditions in SFHA and determine methods to meet the safety objectives in SFHA.
2. To demonstrate the capability to satisfy the qualitative and quantitative requirements of the failure conditions in FHA according to the preliminary design data and information.
3. To evaluate the qualitative and quantitative safety requirements of the system at different levels. Generally, these requirements will be included in documents such as the Purchaser Technical Specification.

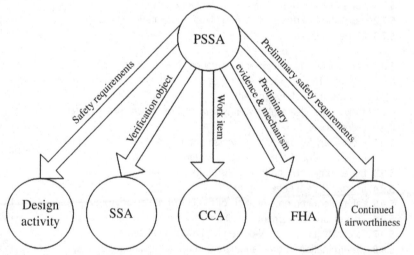

Figure 5.1 Correlation between PSSA and other activities. *PSSA*, Preliminary System Safety Assessment.

4. To evaluate the safety requirements of related activities, such as installation, maintenance, and operation
5. To determine the proposed architecture to satisfy the safety requirements.
6. To determine safety requirements of the interactions with other systems, etc.
7. To produce the list of assumptions on independence used in Fault Tree Analysis (FTA) for ease of validation.
8. To determine the inputs of CMA, etc.

5.2 DETERMINATION OF THE ANALYSIS DEPTH

The analysis method adopted by PSSA may vary from different failure conditions. Whether qualitative and/or quantitative analysis methods are adopted will be determined by a comprehensive analysis of the classification of the failure conditions, the complexity of the systems, and the service experience of similar systems. The criterion for analysis according to FAA AC25.1309-1B (Arsenal) [1] is as follows:

It is acceptable that only the design and installation appraisals are used to verify the safety objectives of "No safety effect" or "Minor" failure conditions. For "Major," "Hazardous," or "Catastrophic" failure conditions, if the design attributes are similar to the existing systems, compliance with the requirements can be demonstrated by verifying the similarity. Otherwise, the verification should be performed in the quantitative method (the quantitative Failure Modes and Effects Analysis (FMEA), FTA, Dependence Diagram (DD), Markov Analysis (MA), and other methods) unless it is a simple or conventional system (Fig. 5.2).

Note: Similarity can be claimed only if the two systems/items have similar functions and classifications of failure conditions, which are performed in equivalent environment with similar usage. Therefore, the system similarity demonstration may be accepted by the authorities for aircrafts from the same family, such as Boeing B737-NG. Otherwise, even if they use the same system products from the same supplier, their similarity usually cannot be accepted due to the different installation and maintenance environments.

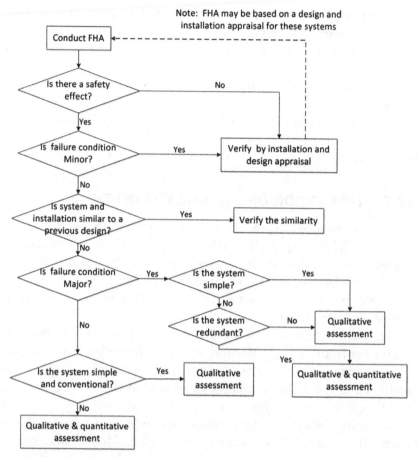

Figure 5.2 Analysis depth of safety assessment.

5.3 ASSUMPTIONS IN PRELIMINARY SYSTEM SAFETY ASSESSMENT

In the process of conducting PSSA, various assumptions may be used or produced. These assumptions need not only further validation but also effective management to ensure their completeness, correctness, and traceability. Assumptions used or produced in PSSA mainly focus on the independence of AND gates in fault trees, average flight duration, failure distribution of different components and assumptions in the design, etc. A brief introduction to the major assumptions mentioned above is provided in the next section to aid in understanding PSSA.

5.3.1 Assumptions of Failure Distribution Types

Before conducting FTA in PSSA, the failures of components are generally assumed to adhere to a bathtub curve, as shown in Fig. 5.3. It can be seen from the figure that the change of the failure rate $\lambda(t)$ can be roughly divided into three stages:

1. *Early Failure Stage*: It appears in the early "infant mortality" stages of the operation of the products. It is characterized by high failure rates in the beginning; however, with the passage of time, the early failures are eliminated, and the failure rate drops rapidly.
2. *Random Failure Stage*: During this stage, there is a lower failure rate and longer duration. It is characterized by an approximately constant failure rate.
3. *Wear-Out Failure Stage*: It appears after a long working period of the components. It is characterized by a rapidly rising failure rate with the passage of time.

Therefore, before new types of components are used on the aircraft, we usually eliminate the early failures via reliability tests, environmental stress screenings, or other methods to maintain constant failure rates for the components that have been installed. We also usually replace components or utilize components with the same life cycle as the aircraft to avoid the increase of failure rates during the wear-out failure stage.

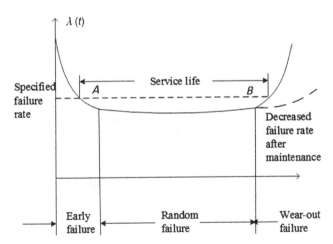

Figure 5.3 Bathtub curve.

In general, when conducting the calculation of PSSA, we assume that $\lambda(t)$ is a constant value in the random failure stage, which approximately corresponds to the value at the bottom of the bathtub curve.

5.3.2 Typical Average Flight Duration

While conducting the PSSA quantitative analysis, the parameter "average flight duration" should be determined. Average flight duration vary depending on the type of aircraft and mission profile. An accurate value should be obtained through market research and statistics combined with previous experience and the users' requirements. Table 5.1 is an example of the average flight duration of different types of aircraft.

5.3.3 Independence Assumptions of AND Gates in Fault Trees

Independence assumptions of AND gates in fault trees are the basis for conducting FTA and PSSA analysis. Further validation and verification of independence should be carried out for AND gates related to catastrophic failure conditions, mainly through the methods of FMEA, CMA, PRA, ZSA, etc.

5.3.4 Other Assumptions in Preliminary System Safety Assessment

The assumptions produced in PSSA are of great number and need to be identified and validated one by one in the development process. In addition to the assumptions mentioned above, there are other analytical assumptions, primarily including:

1. Assumptions in the quantitative FTA, e.g., assuming that the software and human operation are completely reliable and that the failure rate is zero.
2. Assumptions of specific products, such as the storage environment of products, etc.
3. Assumptions of the operation environment, including operation height, temperature, humidity, etc.

Table 5.1 Example of the typical flight duration for different types of aircraft

Type of aircraft	Regional aircraft	Narrow-bodied aircraft	Wide-bodied aircraft
Average flight duration	Approximately 1 hour	2–4 hours	More than 4 hours

4. Assumptions produced by Aircraft Functional Hazard Assessment (AFHA), Preliminary Aircraft Safety Assessment (PASA), and SFHA, conducting the corresponding validation activities in PSSA and System Safety Assessment (SSA);

5. Assumptions in design, such as assumptions of failure detection coverage. When the monitoring is included in the fault tree, two assumptions can be adopted [2]:
   1. The monitor provides 100% failure detection coverage of the item performing the functions;
   2. The monitor verification ("scrub") operation verifies that the monitor is fully operational (i.e., the "scrub" operation provides 100% coverage of the monitor).

   Unfortunately, real-life monitors may not provide 100% coverage. The analyst should consider fine-tuning FTA to account for imperfect coverage.

6. Other assumptions made during specific project.

5.4 PRELIMINARY SYSTEM SAFETY ASSESSMENT INPUTS

Before conducting PSSA, the AFHA/CCA process creates an initial set of safety requirements for the development of the aircraft. Similarly, the SFHA/CCA process creates an initial set of safety requirements for the systems. By combining these initial sets of safety requirements, design decisions are made in the PSSA process. Thus, the following inputs are necessary for PSSA:

1. failure conditions and safety requirements (including DAL, failure probability, etc.) identified in AFHA and PASA;
2. failure conditions and related safety requirements identified in SFHA;
3. preliminary CCA;
4. recommended system architecture;
5. interfaces with other systems and their interactive relationships;
6. the functional list of the systems or items.

5.5 PRELIMINARY SYSTEM SAFETY ASSESSMENT PROCESS

PSSA is a top-down approach for conducting systematic assessment of a proposed architecture and its implementation according to classifications of failure conditions in SFHA and outputting the safety requirements of the subsystems/items. While conducting preliminary assessment on the system architecture, CCA should be fully implemented, especially assessment of the redundancy, isolation and independence of the functional implementation of the system design.

It can be seen that the PSSA process mainly includes the following three aspects:

1. analysis of the safety requirements of the system;
2. assessment of the failure conditions;
3. allocation of safety requirements to the lower level design.

Note: With the development of the formal and simulation technology, formal model-based safety analysis can be a safety analysis method used in PSSA. Detailed information will be given in Chapter 10, Formal Model—Based Safety Analysis Methods and the Application.

5.5.1 Analysis of the Safety Requirements of the System

All safety requirements of the system should be identified from the input data and deemed as the basis of the PSSA process. Safety requirements usually focus on catastrophic, hazardous, and major failure conditions.

During the system architecture and item development, the implementation or integration of each function may result in the appearance of new functions, thus creating new failure conditions. These new failure conditions should be delivered to the FHA process for further assessment. The implementation of various functions may also derive new requirements (e.g., isolation requirements and operational requirements). These new requirements and failure conditions may require additional analysis in PSSA or FHA for validation and verification.

Note: In recent years, aircraft safety issue on the Single Event Effects (SEE) has been drawing a lot of attention. In certain number of type certification, the safety assessment for the SEE has been required by national authorities. Detailed information will be given in Chapter 9, Single Event Effects in Avionics.

5.5.2 Preliminary Assessment of Failure Conditions

For the safety assessment of each significant failure condition (including catastrophic, hazardous, and major failure conditions) in SFHA, PSSA usually adopts FTA as the analysis tool. DD or MA can also be used, but

FTA is more favorable to the assessment at the current stage. Assessment processes that adopt the FTA method mainly complete two major tasks:

1. Establishment of the fault tree for each failure condition; Section 5.7 will provide a detailed description.
2. Preliminary assessment as to whether the system design/architecture meets the safety requirements.

Preliminary safety assessment of the system architecture should be conducted according to the fault trees of each significant failure condition. Assessment includes the following:

1. To demonstrate that how combinations of equipment failures lead to failure conditions.
2. To ensure the independence of the AND gates in the fault trees, the following aspects should be identified:
   a. all separation/isolation requirements, as well as the relevant requirements in CCA;
   b. verification of the independence through tests (ground and flight tests) and/or analysis (such as Common Cause Analysis, CCA);
3. To demonstrate that the proposed system architecture and failure probability budget can satisfy the qualitative and quantitative requirements of relevant failure conditions.
4. To determine the maximum interval of the maintenance task for the hidden failures in the fault trees.
5. To determine the DALs of the related items.
6. To assess significant derived requirements generated in the development of items (if applicable).
7. Incorrect operation.

In the process of system design, implementation of the PSSA process may not produce detailed item level design data. Therefore, assessment of failure conditions in PSSA has to be partly dependent on engineering judgment and operation experience with regard to aircraft types with similar designs. This process will be iterative and constantly improved in the development process.

5.5.3 Safety Requirements Allocated to Lower Level Designs

Each system level safety requirement should be allocated to the items that compose the system. The allocated requirements include:

1. the safety requirements (qualitative or quantitative) allocated to all items (software/hardware);

2. the requirements for installation and design (separation, protection, etc.);
3. the DALs for airborne hardware/software;
4. the safety quantitative requirements;
5. the safety maintenance task and the maximum maintenance interval.

The undesired events and relevant probabilistic budget determined in the system level fault trees are the basis for the detailed design of lower level items.

5.6 PRELIMINARY SYSTEM SAFETY ASSESSMENT OUTPUTS

5.6.1 Output Documents

The results of PSSA should be documented to establish the traceability of the PSSA analysis process. The information that deserves to be maintained includes:

1. compliance methods with safety objectives and requirements in FHA;
2. the updated FHA;
3. materials that support classifications of failure conditions;
4. list of the failure conditions;
5. lower level safety requirements (including DAL);
6. qualitative FTA;
7. preliminary CCA;
8. operational requirements.

5.6.2 Outputs to Lower Level Preliminary System Safety Assessment

PSSA may also be applied for subsystems or items. The inputs of lower level PSSA are the related failure effects, qualitative requirements, probability budget, and DAL determined in the upper level of FHA/PSSA. After gaining the output from the upper level, lower level PSSA of subsystems can be done via the process in Section 5.5.

5.7 PRELIMINARY SYSTEM SAFETY ASSESSMENT AND FAULT TREE ANALYSIS

5.7.1 Roles of Fault Tree Analysis in Safety Assessment

FTA can serve as an effective measure to investigate failure reasons after a major failure or accident occurs; it can be used as a guidance for fault diagnosis and to improve usage scenarios and maintenance plans; it can also be used to spot reliability and safety weaknesses and take measures to improve them.

FTA's graphical representation is hierarchical and named according to its branches. It has strong readability and is easy to understand, which makes FTA a useful tool to conduct safety design by industry and certification authorities. In the process of safety assessment, FTA has the following functions:

1. analyze failure reasons of the top events combined with the system architecture;
2. quantify the probabilities of the top events;
3. allocate the safety requirements of the top events to the lower level events;
4. assess the effects of the development errors through the combination of qualitative and quantitative methods;
5. assess the effects of single and combined failures;
6. assess the effects of the exposure time of the hidden failures on the system safety;
7. assess the source of common cause failures;
8. assess the nature of fail-safe design (fault tolerance and error tolerance);
9. assess the effects of design change on safety;
10. compared with other safety analysis methods, FTA is used most widely in the aeronautical industry.

FTA is done in the process of PASA/ASA and PSSA/SSA.

In the process of PASA, FTA is used to determine the failure reasons of failure conditions in AFHA. The top events of fault trees are the failure conditions in AFHA, and the basic events are usually the failure conditions in SFHA.

In the process of PSSA, FTA is used to allocate the safety requirements of failure conditions identified in SFHA down to lower level items, combining the proposed system architecture and the CCA results.

The information obtained in the detailed design may cause changes in the fault trees. Therefore, in the SSA process, the failure rates from FMES or others will correspond with the basic events of the fault trees, and the top event calculated is the probability of the failure conditions identified in SFHA to verify that the system design meets the safety objectives.

Moreover, problems exposed during the prototype tests and flight tests might lead to changes in hardware or software, as well as changes in the fault trees, and thus the final fault trees will be regarded as part of the safety assessment document.

5.7.2 Calculation of the Fault Tree Analysis's Quantitative Probability

The quantitative analysis of fault trees must be based on a system mission time. During the process of safety analysis, the top event probability refers to the probability of events occurring during flight. Therefore, the system mission time is usually settled as the average flight duration, which should be defined before the calculation. Then, follow the below steps to calculate the quantitative probability for the FTA.

Note: Usually, the fault trees conducted during PSSA and SSA are quite "large" and include inevitably some unquantifiable information, such as development errors. Therefore, before starting quantitative analysis, it is necessary for analysts to deal with the information and form a smaller fault tree by deleting them or settling the failure rate to zero.

1. Determining the minimal cut sets of the fault tree
   The quantitative FTA is based on the minimal cut sets. The minimal cut sets are the combinations of independent basic events. Therefore, it is necessary to determine the data model of the basic events.
2. Collecting the failure rate of basic events
   The failure rate data of basic events in the fault trees can normally be obtained through FMEA/FMES or a corporate database (if applicable) that is provided by suppliers or from the similar products. Data acquisition methods will be instructed in Chapter 7 Failure Modes and Effects Analysis.
3. Determining the risk time and exposure time of basic events
   Risk time designates the period of time within the flight during which an item must fail in order to cause the studied failure effect. Analysts should also confirm the risk time to calculate the occurrence probability of the events after determining the failure rate of basic events:
   - On the condition that a function of an item is used throughout the entire flight, the risk time of the basic event is equal to the average flight duration.
   - On the condition that a function of an item is only used in a particular flight phase, the risk time may start from the preflight test or function launch to the end of the phase in question but is not equal to the average flight duration.

Exposure time should be taken into account when a hidden failure exists. It designates time intervals between two dates for which we know that the hidden failure does not occur (it may be inspections for a maintenance task or preflight check, etc.).

4. Executing numerical computation of the FTA

The occurrence probability of the top event is

$$P(TOP) = P(G_1 + G_2 + \cdots + G_n)$$

$$= \sum_{i=1}^{n} P(G_i) - \sum_{1 \le i < j \le n} P(G_i G_j)$$

$$+ \sum_{1 \le i < j < k \le n} P(G_i G_j G_K) + \cdots + (-1)^{n-1} P(G_1 G_2 \cdots G_n)$$

$$(5.1)$$

$G_i(i = 1, 2, \ldots n)$ is the ith minimal cut set.

In engineering, the minimal cut sets generally contain many basic events. The second term, as well as other terms following it, can be ignored when compared with the first term to calculate in a simpler manner. Therefore, the last-written formula may approximately become:

$$P(TOP) = P(G_1 + G_2 + \cdots + G_n) \approx \sum_{i=1}^{n} P(G_i) \qquad (5.2)$$

The calculation in this book uses the engineering approximation algorithm above.

The occurrence probability of the top event is equal to the sum of probabilities of the minimal cut sets. Hence, the key point in the quantitative equation of FTA lies in calculating the probability of each minimal cut set.

5. Output of the analysis results

Following the above steps, users could compare the occurrence probability of the top event with the safety requirements. If the failure condition is hazardous, and the quantitative safety requirement is 10^{-7} per flight hour, then the occurrence probability of this failure condition must be lower than 10^{-7} per flight hour. Furthermore, the analysis can be finished if the safety requirement is fully met; otherwise, the users should improve the reliability of the components or parts, or even change the design to meet the safety requirement.

5.7.3 Hidden Failures

5.7.3.1 Identification of Hidden Failures

Hidden failures refer to those failures that are latent until they are made known to the flight crew or maintenance personnel. A hidden failure, on its own, cannot lead to accidents directly, but it may loss protection mechanisms of the aircraft or decrease its safety margin, which can increase the risk of subsequent failure conditions. To summarize, hidden failures merely impact the functions that do not support the normal operation but do provide safety margins and abnormal protection to the aircraft. We only can discover it through periodic verification tests, maintenance check, or the power-on tests of the monitor.

There is a checking interval for each hidden failures, equivalence to the exposure time. The interval is normally much longer than the flight duration for maintenance considerations. The key of managing the risk of hidden failures lies in addressing them correctly and strictly limiting the exposure time by maintenance tasks.

5.7.3.2 Calculation of Hidden Failures

In the process of the safety assessment, the basic events of the minimal cut set might contain hidden failures, which will lead to much more complicated calculation. The current commercial analysis software either lacks the ability to handle hidden failures or has methods that are not accurate for aviation areas. Thus, we use it only in the conservative calculation. It is feasible as long as the conservative calculation can meet the safety objectives and have certain safety margins. If not, then the accurate calculation for the minimal cut sets could be used to try to meet the safety objectives.

Based on that fact, the probability calculation of the minimal cut set can be divided into three cases: (1) no hidden failures exist; (2) the conservative calculation is used when hidden failures exist; (3) the more accurate calculation is used when hidden failures exist.

The calculation methods are based on the assumptions as follows:

1. Component/part life length conforms to an exponential distribution and the checked component or part is as good as new after a maintenance check.
2. If $\lambda t \leq 0.1$, the calculation of exponential distribution function can be simplified to $P = \lambda t$, where λ is the failure rate of the component or part and t is the risk time or the exposure time.
3. Applicable to the usual flights, and not associated to specific flights (such as ETOPS and MEL)

4. Maintenance checking interval associated to the hidden failures and the average flight duration are synchronized.
5. A cut-set list has been truncated by limiting the order and/or the acceptable probability of occurrence.

5.7.3.2.1 All Basic Events of the Minimal Cut Set Are Active Failures

In this case, we can directly conduct the quantitative analysis for the fault trees, which is a simple multiplication of the probabilities of each basic event in the minimal cut set. Fig. 5.4 gives a simple demonstration to explain the calculation in this case.

There are two minimal cut sets of the fault tree, AB and CD. The four basic events (A, B, C, D) are active failures. The failure rates and risk times are λ_A, λ_B, λ_C, λ_D and T_{Ra}, T_{Rb}, T_{Rc}, T_{Rd}, respectively. T_0 is the average flight duration

$$
\begin{aligned}
P(TOP) &= P(GATE1 + GATE2) \\
&= P(GATE1) + P(GATE2) \\
&= P(AB) + P(CD) \\
&= P(A)P(B) + P(C)P(D) \\
&= \lambda_A \lambda_B T_{Ra} T_{Rb} + \lambda_C \lambda_D T_{Rc} T_{Rd}
\end{aligned}
\tag{5.3}
$$

λ_A, λ_B, λ_C, λ_D the failure rates of A, B, C, and D.

T_{Ra}, T_{Rb}, T_{Rc}, T_{Rd}, the risk time of A, B, C, and D, the most of case, equivelant to T_0;

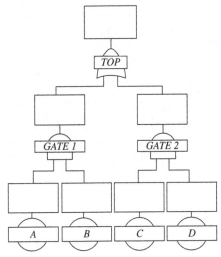

Figure 5.4 Calculation demonstration in the case that basic events of the minimal cut set are all active failures.

The average failure probability of each flight hour is

$$\bar{P}_{fh} = \frac{P(TOP)}{T_0} = \frac{(\lambda_A \lambda_B T_{Ra} T_{Rb} + \lambda_C \lambda_D T_{Rc} T_{Rd})}{T_0} \qquad (5.4)$$

T_0 is the average flight duration

In conservative calculation, analysts can also replace T_R with T_0. The advantage of this approach is that analysts can reduce their workload due to an undetermined risk time.

5.7.3.2.2 The Conservative Calculation When Hidden Failures Exist in the Minimal Cut Set

In this case, the method is similar to the above one, with the only difference being that it regards the exposure time of the hidden failure as the risk time of the basic event. Fig. 5.5 gives a simple demonstration to explain the calculation in this case.

There are two minimal cut sets of the fault tree, AB and CD. The failure rates are λ_A, λ_B, λ_C, and λ_D. The basic events of A, C, and D are active failures, and the risk times are T_{Ra} T_{Rc} and T_{Rd}, while B is a hidden failure with the exposure time T_B.

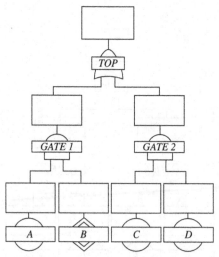

Figure 5.5 The conservative calculation demonstration in the case that hidden failures exist in the minimal cut sets.

$$
\begin{aligned}
P(TOP) &= P(GATE1 + GATE2) \\
&= P(GATE1) + P(GATE2) \\
&= P(AB) + P(CD) \\
&= P(A)P(B) + P(C)P(D) \\
&= \lambda_A \lambda_B T_{Ra} T_B + \lambda_C \lambda_D T_{Rc} T_{Rd}
\end{aligned}
\tag{5.5}
$$

λ_A, λ_B, λ_C, λ_D the failure rates of A, B, C, and D.

T_{Ra}, T_{Rc}, T_{Rd}, the risk time of A, C, and D;

T_B, the expousre time of B

The average failure probability of each flight hour is

$$
\bar{P}_{fh} = \frac{P(TOP)}{T_0} = \frac{(\lambda_A \lambda_B T_{Ra} T_B + \lambda_C \lambda_D T_{Rc} T_{Rd})}{T_0}
\tag{5.6}
$$

It is unnecessary to take the accurate method as long as this conservative calculation can meet the safety objectives and have some safety margins.

5.7.3.2.3 The Accurate Calculation When Hidden Failures Exist in the Minimal Cut Set

A hidden failure may occur in any flight during its maintenance interval. However, the failure condition would not occur until the failures of the rest events in the minimal cut set occurred.

The accurate calculation process is as follows: Firstly, think out each flight case that the hidden failure may occur during the maintenance interval and calculate the failure probability of each case; secondly, add the probabilities of all the cases up and calculate the average value, which is the average failure probability of each flight. The detailed demonstration and deduction is given in Appendix D of SAE ARP4761, and the conclusion can be applied to the case of Fig. 5.5:

$$
\begin{aligned}
P(TOP) &= P(GATE1 + GATE2) \\
&= P(GATE1) + P(GATE2) \\
&= P(AB) + P(CD) \\
&= P(A)P(B) + P(C)P(D) \\
&= \frac{1}{2} \lambda_A \lambda_B T_0 (T_B + T_0) + \lambda_C \lambda_D T_{Rc} T_{Rd}
\end{aligned}
\tag{5.7}
$$

The average failure probability of each flight hour is:

$$\bar{P}_{fh} = \frac{P(TOP)}{T_0} = \frac{1}{2}\lambda_A\lambda_B(T_B + T_0) + \frac{\lambda_C\lambda_D T_{Rc} T_{Rd}}{T_0} \qquad (5.8)$$

It can be found by comparison that the probability difference between accurate and conservative calculation of one hidden failure in the minimal cut set is about $\lambda_A\lambda_B T_B/2$. The difference value will be greater if more hidden failures exist in the cut set. For instance, if there are three hidden failures in the cut set, the conservative results may be even about hundreds of times more than the accurate one depending on the maintenance interval.

In addition, the calculation results will have a significant difference if the hidden failures in the minimal cut set aren't identified correctly. For instance, while accurately calculating the minimal cut set, which contains an active failure and a hidden failure, the average failure probability of each flight is $\frac{1}{2}\lambda_A\lambda_B T_0(T_0 + T_B)$; however, the value becomes $\lambda_A\lambda_B T_0^2$ on the condition that the hidden failure is retreated as active. Obviously, the probability of the later one is much less than the actual value. This leads to incorrect conclusion.

> Note: In the demonstration above, the minimal cut set contains only two basic events. However, in a real analysis process, the minimal cut sets may contain a number of basic events. If there are two or more hidden failures in the cut set, the principle of hidden failure calculation is still applicable, but the equations are more complicated.
>
> If the more accurate calculation is required, the sequence of failures, the period with feared repercussions, and other factors should be taken into account to make the result more precise.

5.7.3.3 Candidate Certification Maintenance Requirement

CCMRs are also some additional contents related to safety assessment of civil aircraft. It comes from the PSSA process. CCMR is in fact of selecting the hidden failures that could lead to hazardous or catastrophic failure conditions while combining with other specific failures or events during the PSSA process.

The calculation methods of the CCMR item check interval is based on quantitative FTA in PSSA. The difference is that the objective of quantitative FTA is to calculate the failure probability of the top events, while the safety objective of the failure conditions will be regarded as the failure probability to calculate the check interval maximum of CCMR items among the basic events. If more than one hidden failure is involved in the minimum cut set, constraint conditions and simultaneous equations could be used to solve the problem.

5.8 DEVELOPMENT ASSURANCE LEVEL ASSIGNMENT

The failure condition of aircrafts may result from random physical failures of the hardware or errors in the process of development. Random physical hardware failure is restricted through probability calculation in quantitative way. Meanwhile, through the development assurance process, system development is ensured to be conducted in an adequately disciplined manner to limit the occurrence possibilities of development errors, which can impact the safety of aircrafts.

Note: The requirements and design errors that occur during the development process will cause unexpected impacts on the airborne high-integrity and complex system and aircraft level function. Nevertheless, it is impractical in real world to perform the exhaustive tests to prove that no error exists. Because of the uncertainty of these errors and the lack of appropriate assessment methods, the development assurance process is taken to minimize the errors in requirements, design and implementation.

For various development errors causing different severities of failure conditions, it is required to conduct different stringencies of development assurance. DAL is the measure of rigor applied to the development process to limit, to a level acceptable for safety, the likelihood of errors during the development process of aircraft/system functions and items that have adverse safety effects.

DAL is not merely applicable not only to the development process of the functions or items but also applicable to the development of interfaces with other correlative functions and items because they may affect the function or item being examined.

5.8.1 Principles for Development Assurance Level Assignment

The DAL assignment is dependent on the classification of the failure condition, as well as the independence, which could limit the impact of errors during the process of development. The more severe the effect of the failure condition is, the higher DAL must be.

According to SAE ARP4754A [3], combined with the severity classification of the failure condition, principles for the DAL assignment are as follows:

1. For the catastrophic failure condition, the assignment principle is:
   - If a catastrophic failure condition could result from a possible development error in an aircraft/system function or item, then the associated Development Assurance process is assigned level A.
   - If a catastrophic failure condition could result from a combination of possible development errors between two or more independently developed aircraft/system functions or items, then either one development assurance process is assigned level A or two development assurance processes are assigned at least level B. The other independently developed aircraft/system functions or items are assigned no lower than DAL C. The development assurance process establishing that the two or more independently developed aircraft/system functions or items are in fact independent should remain level A.

2. For the hazardous failure condition, the assignment principle is:
   - If a hazardous failure condition could result from a possible development error in an aircraft/system function or item, then the associated development assurance process is assigned at least level B.
   - If a hazardous failure condition could result from a combination of possible development errors between two or more independently developed aircraft/system functions or items, then either one development assurance process is assigned at least level B or two development assurance processes are assigned at least level C. The other independently developed aircraft/system functions or items are assigned no lower than DAL D. The development assurance process establishing that two or more independently developed aircraft/system functions or items are in fact independent should remain level B.

3. For the major failure condition, the assignment principle is:
   - If a major failure condition could result from a possible development error in an aircraft/system function or item, then the associated development assurance process is assigned a level C.

- If a major failure condition could result from a combination of possible development errors between two or more independently developed aircraft/system functions or items, then one development assurance process is assigned at least level C or two development assurance processes are assigned at least level D. The development assurance process establishing that the two or more independently developed aircraft/system functions or items are in fact independent should remain level C.

4. For the minor failure condition, the assignment principle is:
   - If a minor failure condition could result from a possible development error in an aircraft/system function or item, then the associated development assurance process is assigned at least level D.
   - If a minor failure condition could result from a combination of possible development errors between two or more independently developed aircraft/system functions or items, then one development assurance process is assigned at least level D.

5.8.2 Function Development Assurance Level and Item Development Assurance Level Assignment

The development process of the aircraft and system can be divided into two phases: the function development phase and item development phase.

1. *Function Development Phase:* The function requirements are confirmed and assigned to every item in this phase. The development process of requirements includes the validation of the requirements set (i.e., the assurance of function integrity and accuracy). It is the Function Development Assurance Level (FDAL) that confirms the level of rigor of the development process of function requirements.

2. *Item Development Phase:* The item (electronic hardware or software) development is accomplished in this phase. The level of rigor of the item development process is confirmed by the electronic hardware or software assurance level (hereinafter referred to as IDAL). In accordance with the corresponding IDAL, the guidelines to objectives required for electronic hardware are included in DO-254, and the guidelines to software objectives are included in DO-178. It should be noted that the boundaries between systems and items is not completely consistent with that between Original Equipment Manufacturers and suppliers or that between suppliers and subsuppliers.

Table 5.2 Top-level function FDAL assignment

Top-level failure condition severity classification	Associated top-level function FDAL assignment
Catastrophic	A
Hazardous	B
Major	C
Minor	D
No safety effect	E

The failure conditions are systematically determined by AFHA and SFHA, which acts as the basis of DAL assignment. Based on AFHA, the scheme of the aircraft function assignment is assessed regarding potential failure conditions to confirm the aircraft level architectures and derived safety requirements for each system contributing to the aircraft level functionality.

DALs are assigned to the aircraft/system functions and items; thus, through appropriate validation and verification processes, the errors occurring in the development process can be reduced as much as possible.

5.8.2.1 Top-Level Function Development Assurance Level Assignment

During the DAL assignment, the FDAL should be assigned to the top-level function according to the most severe top-level failure condition classification.

5.8.2.2 Development Assurance Level Assignment Without Architecture Considerations

DAL can be directly assigned for all the contents under functions by use of Table 5.2 (i.e., FDAL of all the functions supporting the top-level function and IDAL of all items in the architecture are identical to the FDAL of the top-level Function).

Note: For catastrophic failure conditions, if the mitigation measure for systematic errors is a single Level A development assurance process, the authority may require the applicant to prove completely independent validation and verification activities, approaches, and completion criteria to be involved in this process to ensure that potential development errors causing catastrophic effects have been eliminated or mitigated. In this case, it is necessary to ensure that, in the process, development errors can be detected and corrected

(Continued)

(Continued)
by development assurance processes rather than depending on the architecture independence. But it is taking high risk that the only mitigation of errors is through the A Level development assurance for the functions or items with catastrophic failure conditions. This strategy is usually unacceptable to many national authorities, and the architecture mitigation is highly recommended.

5.8.2.3 Development Assurance Level Assignment With Architecture Considerations

After FDAL assignment to a top-level function according to the severity classification of the top-level failure conditions, the architecture should be examined to confirm the FDAL of these system functions. With the consideration of architecture effects, the "independence" should be mainly taken into account.

With the considerations of system architectures, the Functional Failure Set (FFS) can be used as the system approach of DAL assignment. Each FFS and all its members causing a top-level failure condition are confirmed by the system safety assessment approach. The FFS for a given failure condition is confirmed by qualitative safety assessment.

For FDAL and IDAL assignment, the FFS is equivalent to the minimum cut set of the fault tree, whose members represent potential development errors rather than failures. FFS is used to confirm the member combinations causing each failure condition and assign appropriate rigor to reduce potential errors. One failure condition may have at least one FFS, and each FFS may also contain at least one member.

5.9 CASE STUDY OF THE ELECTRICAL POWER SYSTEM ON PRELIMINARY SYSTEM SAFETY ASSESSMENT PROCESS

The purpose of the study case below is to demonstrate the PSSA process, other than a real-system design. Thus, some details of the design are simplified to be more readable.

5.9.1 Preliminary System Safety Assessment Inputs

5.9.1.1 Preliminary Aircraft Safety Assessment and System-Level Functional Hazard Assessment

Chapter 4, System Functional Hazard Assessment, has already formed the safety objectives for the Electrical Power System (EPS), as shown in Table 4.5. PSSA will analyze and verify the rationality of the preliminary system architecture and allocate the safety requirements to lower level items. We will take the failure condition of EPS "24–FC–3 Total Loss of DC Network" as an example. According to the conclusion of SFHA, it is the catastrophic failure condition, while the safety requirement is 10^{-9}/FH so that the related function is developed as FDAL A.

5.9.1.2 System Functional and Operational Requirements

System functional and operational requirements are the necessary inputs for formulating the system preliminary architecture and completing the PSSA process.

The EPS functional requirements are as follows:

1. The primary functional requirements of the power generation system:
   a. To generate electrical power in accordance with the type of airborne electrical equipment and their electricity consumption, including DC electrical power and AC electrical power.
   b. To maintain electrical power quality within design standards.
   c. To transfer the power between the generators and the distributing system.
   d. To provide the status of the system (such as the power supply system configuration, indication and alarm of voltage and current) to the crew.
   e. To provide maintenance data regarding the system.
2. The primary functional requirements of the electrical distribution system:
   a. The mechanical function, to compose the physical interfaces between the generating system and distributing system.
   b. The commutation function, to configure the network, balance the loads, or switch the loads ON/OFF.
   c. The protection function, to protect the wires.
   d. The management function, to control and monitor the network.
   e. The load management function, to optimize the availability of energy for utilization loads versus the availability of the power sources.
   f. To provide to the crew the status of the system (status of the electrical network, generator loading level, etc.).
   g. To provide maintenance data regarding the system.
3. The assumptive aircraft operational requirements related to the EPS:
   a. The average flight duration of the aircraft is 1.2 hours.
   b. The life-cycle of the aircraft: 50,000 flight hours.

5.9.2 Development of the Generating System Architecture

This section illustrates the safety design process for the EPS. It can be noticed that the complexity and safety of the function are both improved iteratively versus accomplishment level of system design.

5.9.2.1 Design for the Single-Channel Normal Power Supply

Step 1: The EPS could be preliminarily designed with single-channel power supply to implement the functions without considerations of the safety, as shown in Fig. 5.6:

- The AC generator supplies electrical power for the AC BUS.
- The AC BUS provides physical interfaces between AC generator and AC power-consumption systems.

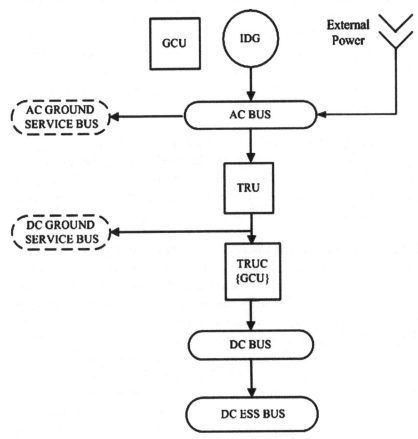

Figure 5.6 Preliminary architecture-1.

- The main function of the GCU is to provide control, monitoring, communication, protection, and self-examination for the power supply channel connected by the GCU.
- The main function of the TRU is to transform three-phase AC power input to DC power for the channel connected by the TRU.
- The DC BUS provides physical interfaces between the TRU and DC power-consumption system.
- The TRUC is controlled by the GCU. When TRU inputs meet the normal operation requirements, the TRU is allowed to be connected to the DC BUS.
- During the operation of the aircraft, an airborne power source providing power for ground equipment and a ground power source providing power for airborne equipment are essential for the maintenance on the ground. we will provide AC and DC GROUND SERVICE BUSes; these two function block diagrams are drawn in the figure with dotted lines.

The single-channel design does not meet the "fail-safe" criteria. In addition, it is impractical to achieve the reliability at 10^{-9}/FH to meet the safety objective. Therefore, the EPS is usually designed with the multichannels.

5.9.2.2 Design for a Dual-Channel Normal Power Supply

Step 2: To improve the safety of the EPS in this case, the dual-channel power supply design is adopted on the aircraft with two engines. Each engine will drive only one channel, and both of the channels can work individually or standby for each other. (Fig. 5.7):

5.9.2.3 Design for the Auxiliary Power Unit Power Supply

Step 3: Taking the aircraft level functional requirements that should be conducted on the ground into account, such as the environment control and related checks, an Auxiliary Power Unit (APU) power supply should be contained in the overall design to support the functions, at the same time, and to backup for the normal power supplies. However, the AGCU has the same design and composition as the GCU so that the common cause failures could easily lead to simultaneous loss of the both devices, and it will take a long time to start the APU, therefore the APU Power Supply is not taken into account in the general safety assessment and FTA. It means that if the device does not work, the aircraft still meets the airworthiness requirements and is allowed to fly continuously. So the APU and AGCU can be the "GO" device in the MEL (Fig. 5.8).

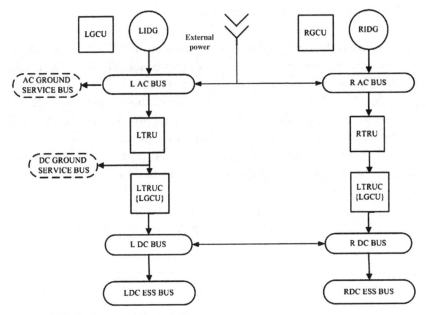

Figure 5.7 Preliminary architecture-2.

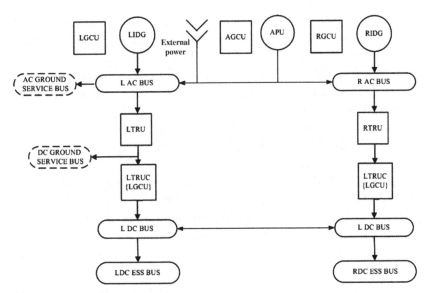

Figure 5.8 Preliminary architecture-3.

5.9.2.4 Design of the Emergency Power Supply

Step 4: "Total Loss of AC Network" and "Total Loss of DC Network" are classified as the catastrophic failure conditions in SFHA. According to the airworthiness requirements, each of them is not allowed to be caused by single failures. However, there are only single AC and DC facilities for generating and distributing power in architecture-1; architecure-2 and architecture-3 contain dual-channels power supply, but the AC and DC facilities in the different channel are same; so a great deal of single failures may occur, resulting in the failures of the DC and AC power supplies onboard. In accordance with the requirements in 25.1351(d) [4]:

FAR25.1351(d)

(d) Operation without normal electrical power. It must be shown by analysis, tests, or both, that the airplane can be operated safely in VFR conditions, for a period of not less than five minutes, with the normal electrical power (electrical power sources excluding the battery) inoperative, with critical type fuel (from the standpoint of flameout and restart capability), and with the airplane initially at the maximum certificated altitude. Parts of the electrical system may remain on if—

1. A single malfunction, including a wire bundle or junction box fire, cannot result in loss of both the part turned off and the part turned on; and

2. The parts turned on are electrically and mechanically isolated from the parts turned off.

To meet the requirement of operating at least 5 minutes without normal power and avoid the Catastrophic and Hazardous failure conditions resulting from the single failures, an RAT generating power channel, which is totally different from an IDG generating power channel, is usually designed to form an independent and dissimilar architecture by the following method:

- The generators are driven by different prime motors:
  - The normal IDG is driven by the engine.
  - The APU generator is driven by the APU.
  - The RAT generator is driven by the ram-air turbine, which has a dissimilar motor electromagnetic design.
- The dissimilarity of the design and composition of GCUs:
  - The GCU and AGCU are controlled by microprocessors.
  - The RAT GCU is controlled by the Programmable Logic Devices (PLD) and analog circuits.
- IDGs and the APU are installed in independent and separate areas, and the Electrical Wire Interconnection System (EWIS) of different channels are installed segregatively.

- The IDG can be operated by the control switch in the cockpit, and the RAT will be operated automatically or manually. There is no operational interface between them.

Because the RAT system is the backup power source only used under emergency circumstances, the failures of the system cannot be correctly examined when the normal power channel works. Although some components will conduct functional tests and self-tests automatically when the power source of the aircraft is started, it's still necessary to determine the operational checking interval by the analysis of MSG-3 and the safety assessment. Relevant failure modes include failures of the RAT mechanism/engine, upper lock of the mechanism, deployment actuator, manual release cable, ADCU and release solenoid.

5.9.2.5 Design for the Battery Power Supply

Step 5: The quality and safety requirements of the power supply also have to be met under the circumstance that the power supply is temporarily suspended, and the quality of the power supply is transiently changed during the conversion course among the normal power channels, the backup power channel, and the emergency power channel. Therefore, adding the main battery and APU battery as continuous power sources not only improves the quality of the power supply but also ensures power distribution when the normal or backup power channels transform into the emergency power channel, which is necessary to ensure the safety of EPS.

Due to the work-time limitation, the batteries are usually not considered in the process of safety assessment, except for the analysis of the conversion process of the emergency power channel (Fig. 5.9).

- Static Inverter (INV): Convert the DC power supply to an AC power supply. The INV should be developed according to the standard of TSO-C73.
- RAT system: The RAT generator produces three-phase AC power, consisting of the deployment actuator, restow pump, uplock latch, RAT GCU, ADCU, manual release cable, and RAT. Under the emergency operation mode, the uplock latch will be automatically deployed by receiving the signal from the ADCU. The RAT can also be deployed by manually releasing the handle in the cockpit.
- RAT GCU: The input source selector control, system protection and self-checking logic will be conducted by means of FPGA logic devices.

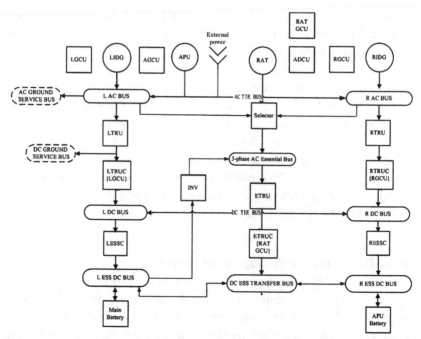

Figure 5.9 Preliminary architecture-4.

- The input source selector: The select power supply for the AC ESS BUS when the RAT is deployed (if there is the failures of the normal AC channel, the RAT will be deployed, and the RAT channel will also automatically be started to provide electrical power to the AC ESS BUS) and the quality of the power generated by the RAT meet the safety requirements.
- Main battery and APU battery: When the TRU cannot provide power, the batteries need to supply electrical power for essential loads and APU startup.

The derived requirements related to the batteries will be produced from the current design, such as the *main battery and APU battery should discharge continually for at least 30 minutes (REF: 24-DSR-1)*. In addition, under the emergency circumstance, to avoid the overload due to the DCESS BUS supplying power to the DC BUS, when batteries supply power to the DC ESSBUS, the derived requirements are needed: *under the emergency cases, the L DC BUS should be isolated from the L DC ESS BUS (REF: 24-DSR-2)*. Therefore, the contactor should be located between both BUSes to avoid using the ESS BUS supply power for the DC BUS.

Because the INV is used as the DC backup power source only under emergency circumstances, the failures of the INV cannot be correctly probed when normal power channels works well. Therefore, it's neccessary to determine the operational check interval by the analysis of MSG-3 and the safety assessment.

5.9.3 Development of the Power Distribution System

When the aircraft is in normal operation, all electric and electronic equipments are eventually powered by the normal power channels. Two IDGs supply power to the AC BUS via the left and right channels to ensure that any failure will only affect the one power supply channel and avoid the transformation of the failures. TRUs convert normal AC power to DC power for the DC BUS. When one normal channel fails, the APU backup channel will replace the failure one to supply power. The APU channel can also be used as the power source for ground maintenance.

When all AC normal power sources fail and the APU cannot work, the batteries and INV will supply power for emergency loads before the RAT generator starts working. When there is no AC power from the generating power system, the RAT should be deployed. There are two ways to deploy the RAT: an automatic deployment signal from ADCU or a manual release handle in cockpit.

The main battery and APU battery supply power for the left and right ESS BUS, respectively. The APU battery is used to start APU and the two batteries will, respectively, supply power for the left and right DCESSBUS as required (Table 5.3).

5.9.4 Interfaces With Other Systems

In terms of its function, the interfaces of the EPS mainly include the cockpit control interfaces, the display and warning interfaces, and other functional interfaces.

- Cockpit control interfaces

    The EPS cockpit overhead panel includes most EPS switches, through which the IDG channel power supply, external AC power supply, and APU generator channel power supply can be controlled. Moreover, the DC channels connected by the batteries can also be controlled by the switches. The panels are given in Table 5.4.

Table 5.3 Bus power source priority

Bus	Bus power source		
	High priority		Low priority
LAC BUS	LIDG	AC TIE BUS	
RAC BUS	RIDG	AC TIE BUS	
AC ESS BUS	RAC BUS	LAC BUS	RAT
AC GROUND SERVICE BUS	LAC BUS	External AC Power	
LDC BUS	LTRU	DC TIE BUS	
RDC BUS	RTRU	DC TIE BUS	
DC GROUND SERVICE BUS	LTRU		
L DC ESS BUS	LDC BUS	DC ET BUS	Main battery
R DC ESS BUS	RDC BUS	DC ET BUS	APU battery
DC ET BUS	ETRU	L DC ESS BUS	R DC ESS BUS

Table 5.4 Examples of cockpit control interfaces of EPS

Switch name	Functional description
LBUS TIE	The control switch of LIDG, supplying power for the AC TIE BUS.
LEFT IDG	LIDG manual disconnect switch
RBUS TIE	The control switch of RIDG supplying power for the AC TIE BUS.
RIGHT IDG	RIDG disconnected switch.
DC BUS TIE	DC connecting logic switch.
ETRU	ETRUC control
Main battery	The battery connecting control.
APU battery	Battery connecting control.
RAT TEST	Provide input for the avionics system to start the RAT BIT.

• Display and alerting interface

The display and alerting interfaces include a lamp light display, cockpit profile displays, or cockpit crew alerting. To provide a visual overview diagram of EPS operation, the system parameters of LGCU,

RGCU, and RAT GCU will be sent to the Data Concentrator Unit (DCU) and the diagram of the power supply will be displayed in Multifunction Displays (MFD).

All EPS Crew Alerting System (CAS) information is conducted by system control logic, such as LGCU and RGCU. Any CAS information produced by the GCU will be sent out through several channels for different purposes. The safety goals for the display and alerting interfaces are usually analyzed in the PSSA of the display and alerting system and sent to the EPS.

- Other functional interfaces

5.9.5 Safety Requirements Analysis of the Electrical Power System

According to the principles described in Section 5.2 and SFHA of the EPS in Table 4.5, due to the high complexity, integrity and lacking of relevant service experience, qualitative and quantitative safety assessments (including FMEA, FTA, and CCA) need to be performed for catastrophic, hazardous, or major failure conditions. In terms of minor or no safety effect failure conditions, only the qualitative analysis is required.

Table 5.5 Examples for interfaces of the EPS with other systems

Interface name	Signal source	Signal destination	Interface description
Weight-on-wheel signal	Landing system	ADCU	ADCU can detect the weight-on-wheel signal to determine whether the aircraft is on the ground or not. The system design ensures that the RAT cannot be deployed when the aircraft is on the ground.
Discrete airspeed signal	DCU	ETRUC	Under emergency circumstances, the RAT supplies power for the aircraft through the ETRU, only when the ETRUC perceive this signal, which is faster than 160 knots.
N2 speed of engine	Engine	GCU	When it is detected that the speed of the engine is faster than the minimum operation speed, the IDG can be started.

Take the following failure condition, e.g., to analyze:

The catastrophic failure condition of "24–FC-3 Total Loss of DC Network," and the safety requirement is located in 10^{-9}/FH.

5.9.6 Preliminary Assessment of Failure Conditions

FTA is done to assess the preliminary architecture for the failure condition "24–FC-3, Total Loss of DC Network" as follows (the definition of the symbols in the FTA can be found in SAE ARP4761):

Split Page-1

Top-event: Total Loss of DC Network.

Description and Analysis:

The DC network contains the L DC BUS, R DC BUS, LDC ESS BUS, RDC ESS BUS, LDC ESSBUS, and DC ET BUS. However, we can see that if the L DC BUS, R DC BUS, and DC ET BUS are lost, the LDC ESS BUS and RDC ESS BUS will not be supplied (the batteries are not taken into account in the safety assessment). In other words, loss of the L DC BUS, R DC BUS, and DC ET BUS will result in the failure conditions 24–FC-3. The loss of each BUS power supply may result from the BUS itself failure or no power transmitted to the BUS (Fig. 5.10).

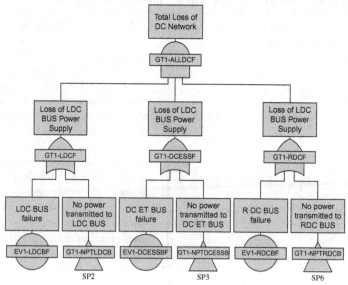

Figure 5.10 FTA split page-1 of the failure condition. *FTA*, Fault Tree Analysis.

Split Page-2

Top-event: No power transmitted to LDC BUS

Description and Analysis:

Both the LTRU and RDC BUS can supply power for the LDC BUS with different priority:

- LTRU can supply power by the LDC channel.
- RDC BUS, which is supplied power by the RTRU, can supply power by the DC TIE BUS (Fig. 5.11).

Split Page-3

Top-event: No power transmitted to DC ET BUS

Description and Analysis:

ETRU, LDC ESS BUS, and RDC ESS BUS can supply power for DC ET BUS with different priority:

- ETRU can supply power by the emergency power channel.
- LDC ESS BUS can supply power by the LDCESS TIE BUS.

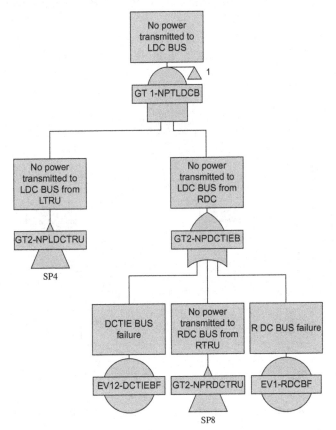

Figure 5.11 FTA split page-2 of the failure condition. *FTA*, Fault Tree Analysis.

- RDC ESS BUS can supply power by the RDCESS TIE BUS (Fig. 5.12).

Split Page –4

Top-event: No power transmitted to LDC BUS from LTRU

Description and Analysis:

There are three direct critical elements which ensure the power can be transmitted to LDC BUS from LTRU: LTRU, LTRUC, and LDC EWIS. Any failures of them will lead to no power transmitted to LDC BUS. Further, we can see that:

- LTRUC is controlled by LGCU, so the failure of LGCU, which may be caused by SEE, can lead to no output from LTRUC.
- The power input of LTRU comes from LAC BUS. If there is no outputs from the LAC BUS, LTRU cannot supply power for LDC BUS (Fig. 5.13).

Split Page-5

Top-event: No power transmitted to DC ET BUS from ETRU

Description and Analysis:

There are three direct critical elements which ensure the power can be transmitted to DC ET BUS from ETRU: ETRU, ETRUC, and

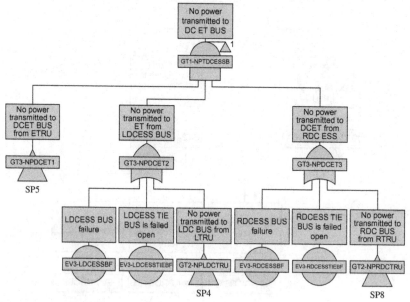

Figure 5.12 FTA split page-3 of the failure condition. *FTA*, Fault Tree Analysis.

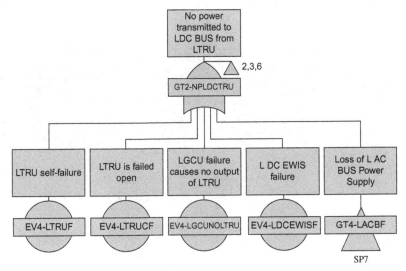

Figure 5.13 FTA split page-4 of the failure condition. *FTA*, Fault Tree Analysis.

Emergency DC EWIS. Any failures of them will lead to no power transmitted to DC ET BUS. Further, we can see that:

- ETRUC is controlled by RATGCU, so the failure of RATGCU can lead to no output from ETRUC.
- The power input of ETRU comes from AC ESS BUS. If there is no output from the AC ESS BUS, ETRU cannot supply power for DC ET BUS.
- There is a switch in the cockpit which controls the ETRU. If the switch or the EWIS between it and ETRU failed, ETRU may stop working.
- Under emergency circumstances, the RAT supplies power for the aircraft through the ETRU, only when the ETRUC perceive discrete airspeed signal, which is faster than 160 knots. If the interface of the signal failed so that the value of speed is less than 160 knots, ETRUC won't allow the power transmitting to the DC ET BUS (Fig. 5.14).

Split Page-6

Top-event: No power transmitted to RDC BUS

Description and Analysis:

Both the RTRU and LDC BUS can supply power for the RDC BUS with different priority:

- RTRU can supply power by the RDC channel.
- LDC BUS, which is supplied power by the LTRU, can supply power by the DC TIE BUS (Fig. 5.15).

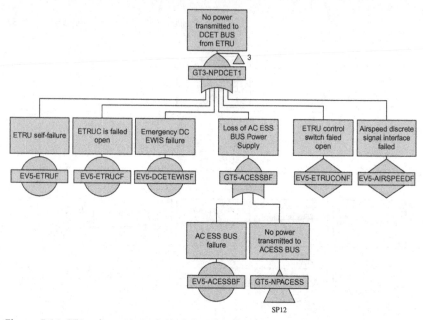

Figure 5.14 FTA split page-5 of the failure condition. *FTA*, Fault Tree Analysis.

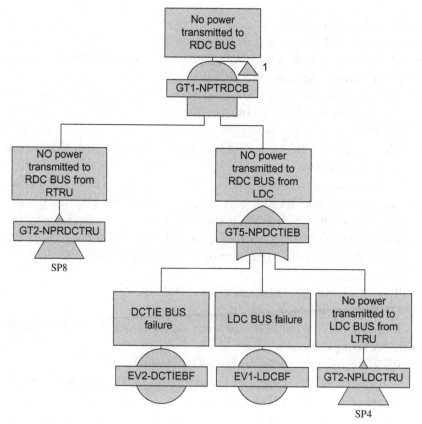

Figure 5.15 FTA split page-6 of the failure condition. *FTA*, Fault Tree Analysis.

Split Page -7

Top-event: Loss of LAC BUS Power Supply

Description and Analysis:

Both the LIDG and RAC BUS can supply power for the LAC BUS with different priority:

- LIDG can supply power by the LAC channel.
- RAC BUS, which is supplied power from the RIDG, can supply power by the AC TIE BUS.
- RBUSTIE and LBUSTIE switches in the cockpit are used to control the AC TIE BUS. Any failure of them may lead to no power from RAC BUS to LAC BUS (Fig. 5.16).

Split Page-8

Top-event: No power transmitted to RDC BUS from RTRU

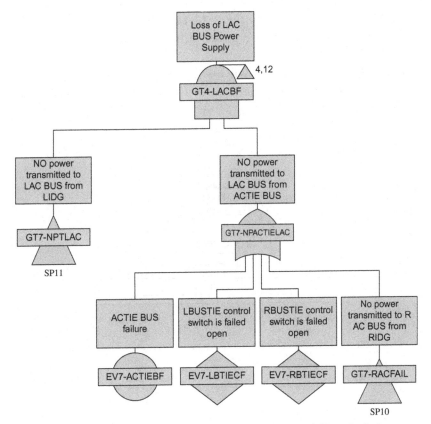

Figure 5.16 FTA split page-7 of the failure condition. *FTA*, Fault Tree Analysis.

Description and Analysis:

There are three direct critical elements which ensure the power can be transmitted to RDC BUS from RTRU: RTRU, RTRUC, and RDC EWIS. Any failures of them will lead to no power transmitted to RDC BUS. Further, we can see that:

- RTRUC is controlled by RGCU, so the failure of RGCU can lead to no output from RTRUC.
- The power input of RTRU comes from RAC BUS. If there is no output from the RAC BUS, RTRU cannot supply power for RDC BUS (Fig. 5.17).

Split Page-9

Top-event: Loss of RAC BUS Power Supply

Description and Analysis:

Both the RIDG and LAC BUS can supply power for the RAC BUS with different priority:

- RIDG can supply power by the RAC channel.
- LAC BUS, which is supplied power from the LIDG, can supply power by the AC TIE BUS.
- RBUSTIE and LBUSTIE switches in the cockpit are used to control the AC TIE BUS. Any failure of them may lead to no power from LAC BUS to RAC BUS (Fig. 5.18).

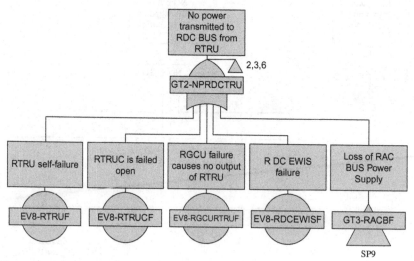

Figure 5.17 FTA split page-8 of the failure condition. *FTA*, Fault Tree Analysis.

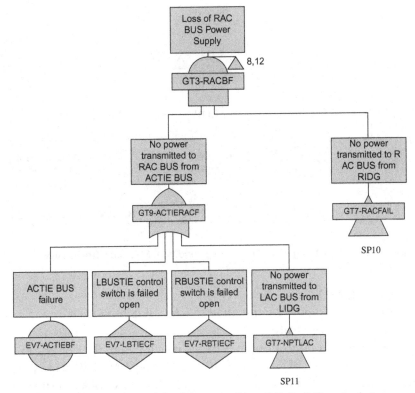

Figure 5.18 FTA split page-9 of the failure condition. *FTA*, Fault Tree Analysis.

Split Page-10

Top-event: No power transmitted to RAC BUS from RIDG

Description and Analysis:

There are two direct critical elements which ensure the power can be transmitted to RAC BUS from RIDG: RIDG and RAC EWIS. Any failures of them will lead to no power transmitted to RAC BUS. Further, we can see that:

- RIDG is controlled by RGCU, so the failure of RGCU can lead to no output from RIDG.
- There is a switch in the cockpit which controls the RIDG. If the switch or the EWIS between it and RIDG fails, RIDG may stop working.
- When it is detected that the engine N2 speed is faster than the minimum operation speed, the RIDG can be started. If the interface fails so that the speed keeps slower than the minimum operation speed, RIDG cannot supply power for RAC BUS (Fig. 5.19).

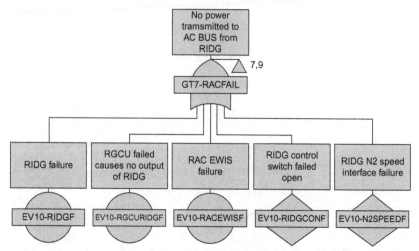

Figure 5.19 FTA split page-10 of the failure condition. *FTA*, Fault Tree Analysis.

Split Page-11

Top-event: No power transmitted to LAC BUS from LIDG

Description and Analysis:

There are two direct critical elements which ensure the power can be transmitted to LAC BUS from LIDG: LIDG and LAC EWIS. Any failures of them will lead to no power transmitted to LAC BUS. Further, we can see that:

- LIDG is controlled by LGCU, so the failure of LGCU can lead to no output from LIDG.
- There is a switch in the cockpit which controls the LIDG. If the switch or the EWIS between it and LIDG fails, LIDG may stop working.
- When it is detected that the engine N2 speed is faster than the minimum operation speed, the LIDG can be started. If the interface fails so that the speed keeps slower than the minimum operation speed, LIDG cannot supply power for RAC BUS (Fig. 5.20).

Split Page-12

Top-event: No power transmitted to ACESS BUS

Description and Analysis:

RAC BUS, LAC BUS, and RAT can supply power for AC ESS BUS with different priority. The selector will choose the power source for the AC ESS BUS according to the current operation situation (Fig. 5.21).

Split Page-13

Top-event: RAT power supply failure

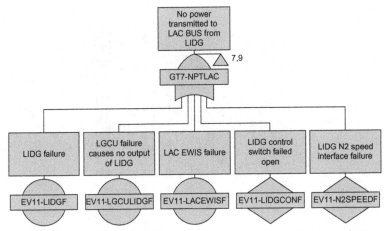

Figure 5.20 FTA split page-11 of the failure condition. *FTA*, Fault Tree Analysis.

Description and Analysis:

There are mainly four factors which can lead to the failures of RAT power supply as follows:

- RAT cannot be deployed.
- DC overload causes the failures of the RAT.
- BIT function fails so that the RAT cannot work.
- No power supplied by the RAT. It can be caused by the failures of RAT mechanism generator, failures of the RAT GCU to control the RAT, and failures of the ADCU (Fig. 5.22).

Split Page-14

Top-event: RAT deployment failure

Description and Analysis:

The deployment of the RAT is completed by the cooperation among mechanical uplock, deployment actuator, and restow pump. Besides, the control command and power for deployment under the emergency situation is necessary input. According to the architecture, the batteries will supply power for the RAT deployment under the emergency situation (Fig. 5.23).

Split Page-15

Top-event: DC overload causes RAT failure

Description and Analysis:

LESSC and RESSC are used to avoid the overload due to the DCESS BUS supplying power to the DC BUS, when batteries supply power to the DC ESS BUS under the emergency situation. If LESSC or RESSC is

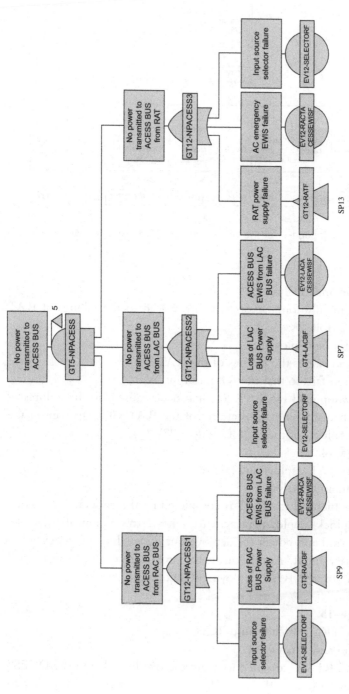

Figure 5.21 FTA split page-12 of the failure condition. *FTA*, Fault Tree Analysis.

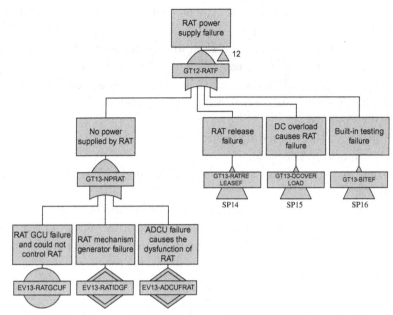

Figure 5.22 FTA split page-13 of the failure condition. *FTA*, Fault Tree Analysis.

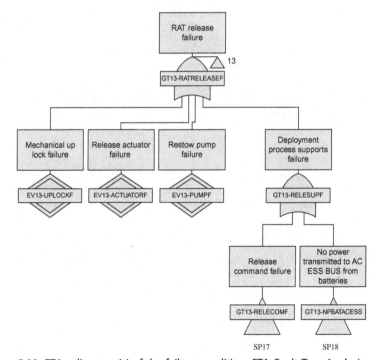

Figure 5.23 FTA split page-14 of the failure condition. *FTA*, Fault Tree Analysis.

failed closed, DC overload will happen and lead to the failures of the RAT. Further, we can see that:

- There are the switches in the cockpit to control the LESSC and RESSC. The failure of the switches will lead to the ESSC failed and closed.
- ESSC is controlled by the TRUC. If TRUC fails to detect or control them, ESSCs may fail and closed too (Fig. 5.24).

Split Page-16

Top-event: Built-in testing failure

Description and Analysis:

There are three factors which can lead to the loss of BIT function of the RAT:

- The BIT function of the RAT is controlled by the RATGCU.
- There is the switch in the cockpit to control the RAT TEST. The failure of the switch will lead to the BIT function doesn't work.

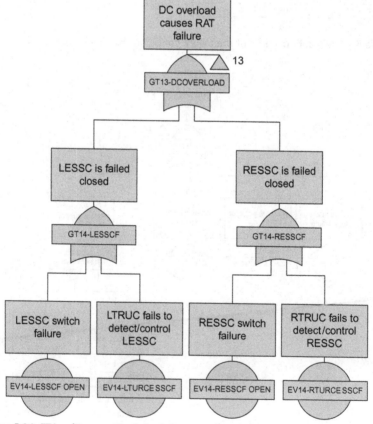

Figure 5.24 FTA split page-15 of the failure condition. *FTA*, Fault Tree Analysis.

- The failures of the avionics BIT signal may lead to the loss of BIT function of the RAT (Fig. 5.25).

Split Page-17

Top-event: deployment command failure

Description and Analysis:

There are two means of deploying the RAT:

- Manual releasing RAT by the handle in the cockpit and release cable.
- Automatic deploying RAT by receiving the signal from the ADCU. The ADCU can detect the weight-on-wheel signal from the landing system to determine whether the aircraft is on the ground or not. If ADCU fails or landing system misleadingly indicates the aircraft is on the ground, the RAT cannot be deployed (Fig. 5.26).

Split Page-18

Top-event: No power transmitted to AC ESS BUS from batteries

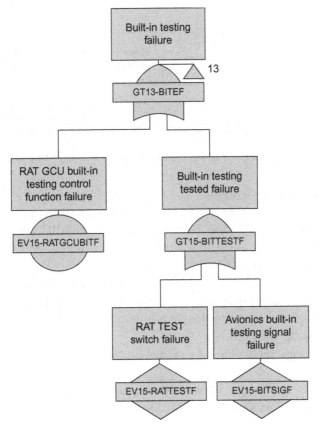

Figure 5.25 FTA split page-16 of the failure condition. *FTA*, Fault Tree Analysis.

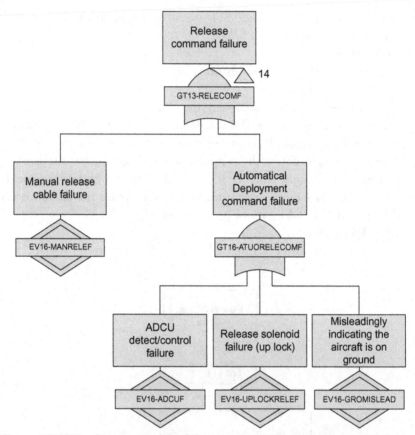

Figure 5.26 FTA split page-17 of the failure condition. *FTA*, Fault Tree Analysis.

Description and Analysis:

At the beginning of the emergency situation, the batteries are required to supply power for AC ESS BUS through LDC ESS BUS, INV, and the EWIS. Main battery and APU battery both can supply power for LDC ESS BUS with different priority (Fig. 5.27).

As for the catastrophic failure condition, while establishing the FTA, the following assumptions will be acquired to ensure the independence between the basic events in the same minimum cut sets, and they will be validated by the SSA process:

- The average flight duration of the aircraft is 1.2 flight hours (REF: 24-ASS-1).
- The EPS can supply adequate power to the loads under normal or emergency situations (REF: 24-ASS-2).
- The failure rate of components of the EPS conforms to the exponential distribution (REF: 24-ASS-3).

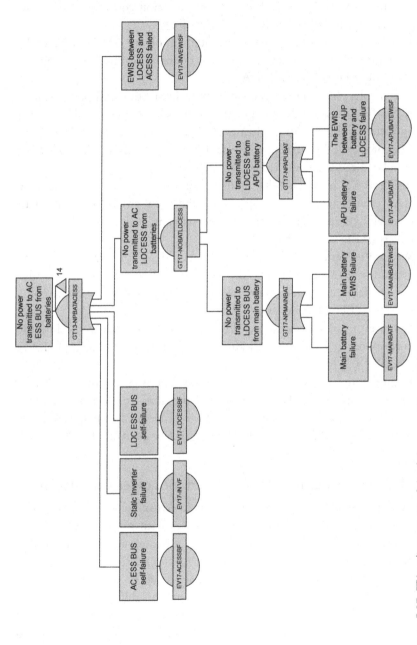

Figure 5.27 FTA split page-18 of the failure condition. *FTA*, Fault Tree Analysis.

- The crew can correctly utilize the human—machine interface components relevant to the EPS (REF: 24-ASS-4).
- The monitor can provide 100% failure detection coverage of the item performing the function (REF: 24-ASS-5).
- The regular functional checking intervals of the INV: 2000 flight hours (the checking intervals in actual operation cannot be more than 2000 flight hours) (REF: 24-ASS-6).
- The regular checking intervals of the release function and power supply of the RAT system: 2000 flight hours (the checking intervals in actual operation will be no more than 2000 flight hours) (REF: 24-ASS-7).

The process of establishing the fault trees will derive the following safety-related requirements at least:

- If the RAT automatic deployment fails while needing to start the RAT emergency power supply, then the crew should be warned to manually release the functions; otherwise, it will cause catastrophic effects (REF: 24-AFM-01).
- When the generator fails, the crew should be warned to close the generator to avoid affecting the safety of the aircraft. It should also prevent false protection of the generator, which will cause catastrophic failures (REF: 24-AFM-02).
- In the event of a BUS failure, the crew should be warned to close the corresponding BUS to avoid affecting the safety of the aircraft (REF: 24-AFM-03).
- In emergency cases, the batteries should supply power to the deployment process of the RAT. Therefore, the abnormal situation can be checked to make sure that the RAT works normally, including overheat (REF: 24-AFM-04) and discharge (REF: 24-AFM-05).

5.9.7 Safety Requirements Allocated for Lower Level Items

Analyze the minimal cut sets of the fault trees to allocate safety requirements for the lower level items and interfaces with other systems. There are approximately 1000 minimal cut sets in this case, and we assume that the safety objective allocated of each one is 1.0E-12/FH to facilitate the allocation at the early phase. But in practice, it can be updated in a more flexible manner by combining the implementation difficulty, the technological ability of suppliers, costs and other factors, as long as the sum of the allocated safety objectives of all the minimal cut sets does not exceed the overall safety objectives of the failure condition. Here is a list of just some of the parts that contain the least members to illustrate (Table 5.6).

Table 5.6 Examples of safety requirements assignment

Number	Minimal cut sets	Safety objective (/FH)	Allocated safety requirements (/FH)
1	EV11-LIDGF	10^{-12}	10^{-4}
	LIDG failure		
	EV10-RIDGF		10^{-4}
	RIDG failure		
	EV13-RATIDGF		10^{-4}
	RAT mechanism/generator failure		
2	EV11-LIDGF	10^{-12}	10^{-4}
	LIDG failure		
	EV10-RIDGF		10^{-4}
	RIDG failure		
	EV16-MANRELEF		10^{-4}
	Manual release cable failure		
3	EV11-LIDGF	10^{-12}	10^{-4}
	LIDG failure		
	EV10-RIDGF		10^{-4}
	RIDGfailure		
	EV16-GROMISLEAD		10^{-4}
	Misleadingly indicating the aircraft is in the ground mode		
4	EV11-LIDGF	10^{-12}	10^{-4}
	LIDG failure		
	EV10-RIDGF		10^{-4}
	RIDG failure		
	EV17-ACESSBF		10^{-4}
	AC ESS BUS self-failure		
5	EV11-LIDGF	10^{-12}	10^{-4}
	LIDG failure		
	EV10-RIDGFRIDG failure		10^{-4}
	EV15-BITSIGF		10^{-4}
	Avionics built-in test signal failure		
6	EV11-LIDGF	10^{-12}	10^{-4}
	LIDG failure		
	EV10-RGCURIDGF		10^{-4}
	RGCU failure causes no output of the RIDG		
	EV13-RATGCUF		10^{-4}
	RATGCU failure causes failure to pass the built-in and functional tests		
7	EV11-LIDGF	10^{-12}	10^{-4}
	LIDG failure		
	EV10-N2SPEEDF		10^{-3}
	Right engine N2 speed interfaces failure		
	EV17-ACESSBF		10^{-5}
	AC ESS BUS self-failure		
8	......	10^{-12}	

As for catastrophic failure conditions (such as this example), the priority is to ensure that there is no common cause failures existing among the basic events of the minimal cut sets. Therefore, after analyzing the above minimal cut sets, the following independence derived requirements can be drawn (including installation, design, maintenance, specific risks, etc.) as the input of CCA analysis:

- The LIDG/RIDG is independent of the RAT (REF: 24-DSR-3).
- The LIDG/RIDG is independent of the RAT manual release cable (considering the aspects of design, installation, specific risks, etc.) (REF: 24-DSR-4).
- The LIDG/RIDG is independent of the Weight-on-Wheel Signal interface with the landing system (REF: 24-DSR-5).
- The LIDG/RIDG is independent of the avionics built-in testing signal (REF: 24-DSR-6).
- The RGCU and RAT GCU are independent (REF: 24-DSR-7).
- The LIDG, right engine N2 speed interface, and AC ESS BUS are independent (REF: 24-DSR-8).
- ......

Thereafter, based on the assumptions that the basic events in the same minimal cut set are independent, the safety requirements can be allocated to the basic events, as shown in Table 5.6. It also can allocate the safety requirements in a flexible manner by combining the implementation difficulty, the technological ability of suppliers, costs, and other factors. However, it must ensure that the sum of the allocated safety requirements of all the minimal cut sets is no more than the overall safety objectives of the failure condition. It is suggested that preliminary verification be conducted with the historical statistics of failure rates for each basic event, reliability test analysis, or relevant reliability standard to ensure that the allocated requirements are practical. For example, according to the present industrial capability, the failure rate of the EWIS is generally 10^{-6}/FH, while the generator has complex structures with a failure probability of 10^{-4}/FH, and that of the digital signal interface components is usually 10^{-4}/FH. Therefore, in the cut set of number 7, it is advised to allocate 10^{-5}/FH to "EV16-ACESSBF AC ESS BUS self-failure" as safety requirements, while "EV11-LIDGF LIDG failure" is allocated for 10^{-4}/FH and "EV10-N2SPEEDF right engine N2 speed interface failure" is allocated for 10^{-3}/FH.

5.9.8 Determination of the Candidate Certification Maintenance Requirements

Summarize the hidden failures related to the failure condition 24–FC–3 as the CCMR items:

1. *EV13-RATIDGF,* RAT mechanism/generator failures; the functional checking interval is assumed to be 2000 flight hours.
2. *EV13-UPLOCKF,* Mechanical uplock failure; the functional checking interval is assumed to be 2000 flight hours.
3. *EV13-ACTUATORF,* Deployment actuator failure; the functional checking interval is assumed to be 2000 flight hours.
4. *EV16-LOCKRELEF,* Release solenoid failure (uplock); the functional checking interval is assumed to be 2000 flight hours.
5. *EV16-MANRELEF,* Manual release cable failure; the functional checking interval is assumed to be 2000 flight hours.
6. *EV13-ADCUFRAT,* ADCU failure causes the dysfunction of RAT; the functional checking interval is assumed to be 2000 flight hours.
7. *EV13-PUMPF,* Restow pump failure; the functional checking interval is assumed to be 2000 flight hours.

More detail information for the confirmation of CCMR items is given in Chapter 8, System Safety Assessment.

5.9.9 Development Assurance Level Assignment

5.9.9.1 Function Development Assurance Level Assignment

24–FC–1, "Total Loss of AC Network," and 24–FC–3, "Total Loss of DC Network," are both catastrophic failure conditions according to the conclusion of SFHA, and therefore the corresponding FDALs of the AC and DC power supply function are level A.

5.9.9.2 Item Development Assurance Level Assignment

In the EPS, the main highly integrated and complex components include the GCU, RATGCU, BPCU, battery controller, ADCU, etc. The GCU and BPCU contain the airborne software. It is necessary to assign IDAL for the airborne complex avionic hardware and software.

- The IDAL assignment of the GCU and RAT GCU

   The FFS is determined through Fig. 5.28. In the normal AC power supply system, neither the left nor the right AC power supply function have functional independence, and the LGCU and RGCU do not

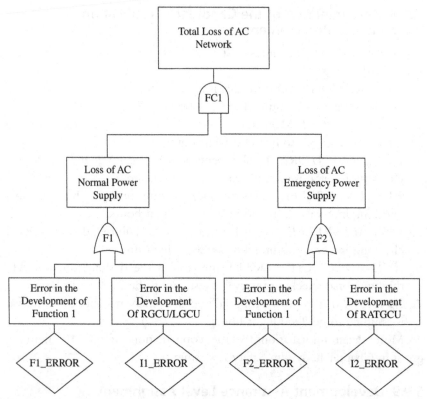

Figure 5.28 Part of the functional failure logic diagram of "Total Loss of AC Network."

have item independence either. Therefore, when assigning DAL for them, they can be considered together and assigned the same FDAL and IDAL. However, normal and emergency power supply channels meet the requirements of functional independence, and the RAT GCU and L/R GCU satisfy the requirement of item independence, so the functional failure logic diagram can be formed as follows:

According to Fig. 5.28, the FC1 failure condition has the following minimum FFS:

- F1 error and F2 error, or
- F1 error and I2 error, or
- I1 error and F2 error, or
- I1 error and I2 error.
- DAL assignment (Table 5.7).

Table 5.7 Examples for the DAL assignment of catastrophic failure conditions

FDAL Assignment		IDAL Assignment		Annotation
F1	F2	I1	I2	
B	B	B	B	Acceptable(final decision in our case)
		A	C	Unacceptable: F1 level B & I2 level C are not allowed, and I2 does not support the assignment of F2 level B
		C	A	Unacceptable: F2 level B & I1 level C are not allowed, and I1 does not support the assignment of F1 level B
A	C	A	C	Acceptable
		C	A	Unacceptable: F2 level C & I1 level C are not allowed, and I1 does not support the assignment of F1 level A
		B	B	Unacceptable: F2 level C & I1 level B are not allowed, and I1 level B does not support the assignment of F1
C	A	A	C	Unacceptable: F1 level C & I2 level C are not allowed, and I2 does not support the assignment of F2 level A
		C	A	Acceptable
		B	B	Unacceptable: F1 level C & I2 level B are not allowed, and I2 level B does not support the assignment of F2

To summarize, we can assign DALs as follows:

- F1 level B & F2 level B & I1 level B & I2 level B;
- F1 level A & F2 level C & I1 level A & I2 level C;
- F1 level C & F2 level A & I1 level C & I2 level A;

However, market research tells us that the first choice costs the least, and a decision can be made at last:

"the FDALs of the normal power supply and emergency power supply are level B, and as for GCU and RAT GCU, the IDAL assigned are level B."

Comprehensively considering all the failure conditions and their classifications in SFHA, it is clear that there is not only a GCU or RAT GCU failure that can cause a catastrophic failure condition. Therefore, it can be concluded that assigning the GCU and RAT GCU as IDAL B level can meet the requirements of each failure condition.

- DAL assignment for BPCU

  Because the failure of the BPCU will cause the Minor failure condition of "losing the AC ground service power supply," the level of software and hardware in the BPCU can be assigned as level C.

- DAL assignment for the battery controller

  The batteries supply power for the deployment of the RAT. If the battery controller fails at the time of conversion between the normal and emergency power channels which results from the normal power channels fail, the RAT cannot be deployed and will have catastrophic effects. But taking into account that the battery system is just used temporarily during the emergency situation, and the normal and emergency channels are independent, the IDAL is determined to be level B at least.

- DAL assignment for ADCU

  ADCU controls the deployment of the RAT. If the ADCU fails at the time of conversion between the normal power channels and emergency power channels which results from the normal power channels fail, the RAT cannot be deployed and will have catastrophic effects. But taking into account that the ADCU is just used during the emergency situation, and the normal and emergency channels are independent, the IDAL is determined to be level B at least.

REFERENCES

[1] AC25.1309-1B (Arsenal). System design and analysis. FAA; 2002.
[2] SAE ARP 4761. Guidelines and methods for conducting the safety assessment process on civil airborne systems and equipment. Society Automotive Engineers; 1996.
[3] SAE ARP 4754A. Guidelines for development of civil aircraft and systems. Society Automotive Engineers; 2010.
[4] FAR25. Airworthiness standards: transport category airplane; 2007.

CHAPTER 6

Common Cause Analysis

Contents

6.1 WHY COMMON CAUSE ANALYSIS?

Satisfying safety objectives may require independence among functions, systems, or items. Therefore, it is necessary to ensure the existence of independence or that the risks related to independence are acceptable. Common Cause Analysis (CCA) provides the tools to verify independence or to identify specific independences. In addition, CCA can identify single failures or external events that can lead to catastrophic or hazardous failure conditions, and it can also capture certain safety requirements to prevent these risks. For instance, to prevent the risks caused by an uncontained turbine engine, it may have the minimum separation requirements for essential systems and their components, as appropriate.

CCA includes Zonal Safety Analysis (ZSA), Particular Risk Analysis (PRA), and Common Mode Analysis (CMA). Common cause resources mainly include hardware or software design errors, software coding errors, compiler errors, requirement errors, environmental factors, hardware failures, installation errors, operational errors, production/repair defects, stress-related events (abnormal), etc. [1].

Civil Aircraft Electrical Power System Safety Assessment
DOI: http://dx.doi.org/10.1016/B978-0-08-100721-1.00006-6
157

The principle of dealing with common causes is that functions whose failure conditions are catastrophic or hazardous will not be affected by the common causes. If affected, all redundant channels of these functions should not be lost, and the remainder of redundant functions can still meet the expected quantitative requirements.

6.2 COMMON MODE ANALYSIS

On June 4, 1996, the maiden flight of the Ariane 5 launcher ended in failure. Only 37 seconds after initiation of the flight sequence, the launcher veered off its flight path, broke up, and exploded. The investigation showed that the accident was caused by software development errors regarding the speed and position calculation-related control computer. The computer system had adopted a redundant design, i.e., two physically isolated and independent channels were controlled by two computers. However, because the hardware and software of the computers in the two channels were identical, a common software development error led to the loss of all functions of the computer system, resulting in a failed launch.

For another example, a redundant design is adopted in the normal power system of an aircraft, which enables the two redundant channels to be independent. However, because of a design error of the system installed in the aircraft, the negative feeders of the two channels were physically fixed in the same, or a very close, location. Once there is a structural damage at this location, it can lead to the feeder circuit being fully interrupted in both the two redundant channels, and it can cause the loss of normal power supply of the aircraft.

Again, a backup channel of a critical airborne system is designed to be monitored by a monitoring function. However, due to the design error, the control circuit implementing the backup function channel is transferred to the monitoring function circuit through a common electrical connector. During flight, when the primary function of the system failed, the pilot found that the backup channel had also failed when using it.

In practice, it is unrealistic to obtain perfect, absolute, and/or theoretical independence from an engineering viewpoint. Design errors, manufacturing errors, maintenance errors, and component failures may affect the design independence, so it is necessary to identify these failure modes that affect independence and to evaluate the extent to these failure modes that affect the aircraft. CMA is a qualitative analysis method to support the evaluation of independence. In CMA, engineering experience is systematically applied to review function, architecture, development, implementation, manufacturing, maintenance, and operation in a logical way. CMA is performed throughout the safety assessment process and may be applied at any indenture levels (aircraft, system or item).

During the "requirements development and validation phase," CMA is performed based on the requirements generated in Preliminary Aircraft Safety Assessment (PASA)/Preliminary System Safety Assessment (PSSA).

During the "verification phase," CMA is performed based on the requirements generated in Aircraft Safety Assessment (ASA)/System Safety Assessment (SSA).

The aircraft level CMA activity describes how the CMA activities support the PASA in ensuring that the independence principles can be satisfied by the proposed design and how the CMA activities verify that the independence principles have been maintained or have not been compromised by the implementation.

Similar to the aircraft level CMA, the system level CMA activity describes how the CMA activities support the PSSA and SSA.

Because this book is about safety assessment of the Electrical Power System (EPS), only the system level CMA activity is described here.

6.2.1 Common Mode Analysis Inputs

CMA is generally aimed at catastrophic and hazardous failure conditions.

The CMA inputs include:
- CMA checklist
- Independence principles and/or independence requirements identified by the PSSA (or PASA at aircraft level)
- Some characteristics of the system, including characteristics relevant to system operation and installation. It mainly includes the following contents:
  - System design architecture and installation plan
  - Characteristics of the equipment and its components
  - Maintenance and test tasks
  - Crew procedures
  - The technical specifications of systems, equipment, software, etc.

When considering the system characteristics, it is necessary to understand some safety precautions that eliminate the common causes, including the following:

- Differences (nonsimilarity, redundancy, etc.) and isolation
- Test and preventive maintenance procedures
- Design control and prevention used in the design process (quality procedures, design reviews, etc.)
- Review of procedures and technical specifications
- Personnel training
- Quality control, etc.

From above, we can sort out the data sources and supporting documents of the input information, including the following:

- CMA checklist
- System requirements documents
- System architecture and design description documents
- System interface description documents
- AFHA report and SFHA report
- System PSSA report

6.2.2 Common Mode Analysis Checklist

1. Generic checklist

To avoid neglecting common cause problems and ensure the smooth proceeding of CMA, it is generally necessary to formulate a generic CMA checklist. The generic checklist includes common cause type, common cause subtype, common cause source, and common cause failures and errors.

The common cause type is the initial classification of failures or errors from design, manufacturing, installation, operation, maintenance, test, calibration, and environmental aspects. Common cause subtype is a further refinement of the common cause type, e.g., the common cause type in design may include architecture, process, technical specifications, etc.

The common cause source is a more specific description of common cause on the basis of common cause subtype. Common cause failures and errors are detailed descriptions of the causes of common cause failures. Common cause source, common cause failures and errors should be based on the existing data and experience, such as the common sense of failure and experiences in similar aircraft.

The level of details of the generic checklist depends on the level of technical research, as well as the complexity and novelty of the system.

2. Tailor the generic checklist

The generic checklist concerns a large number of common cause failures or errors throughout a broad spectrum of designs, which may be addressed across the whole aircraft project. Thus, it is necessary to tailor the generic checklist to suit the particular project under consideration. Project-specific CMA checklists are derived based on the generic checklist, previous experience, and knowledge of object. Checklists are the basis of CMA activities.

6.2.3 Common Mode Analysis Process

The CMA process is summarized as follows: For each independence requirements identified by the PSSA, it is necessary to select the appropriate common cause types and common cause sources based on the CMA checklist. Common cause failure and error analysis is then performed for each given common cause source.

Specifically, first of all, it is required to check the applicability of common cause type and common cause subtype in the checklist to the independence requirements. Once the common cause type and the common cause subtype are selected, the common cause source for this type must be specified. The level of similarity (similarities and differences) of these specific characteristics can be analyzed by referring to the specific characteristics of the relevant content of the independence requirement to identify the common cause source.

Common cause failures and errors are then analyzed and determined for each potential common cause source. For example, a common cause source in a "Loss of electrical power" failure condition is that "if two channels use the same generator, it will result in loss of the electrical power supply simultaneously in two normal channels." The common cause failure and error is that "the same generator leads to the loss of electrical power supply in channel A and channel B."

For each common cause failure and error, it is required to determine the necessary materials to confirm the compliance with independence principles, such as the effects analysis and the detection methods.

For common cause failures and errors that do not comply with the independence requirements, it is necessary to determine whether the common causes are acceptable on consideration, such as design precautions, production, software, etc. If supporting documents are required, relevant analysis should be performed as early as possible and be indexed in the CMA report.

For the unacceptable common cause problems, a non conformance sheet needs to be submitted to the SFHA and PSSA activities, including a detailed description of the nonconforming common causes and its associated recommendations.

6.2.4 Common Mode Analysis Outputs

1. CMA output materials:
   a. Specific CMA checklist
   b. Evaluation of each independence principle
   c. Identified deficiencies that may violate independences
   d. Any assumptions of CMA activities
   e. Reference documents, drawings, and support material used in the analysis
2. The relationship between CMA, ZSA, and PRA

   PRA and ZSA are not part of the CMA. However, the potential common cause effects of these sources cannot be ignored. If a potential common cause is limited to a particular zone or a particular risk, the CMA analyst needs to confirm that the ZSA and PRA include potential common cause problems. Otherwise, the requirement for proof shall be presented to the ZSA analyst or to the PRA analyst.

6.2.5 Case Study of Common Mode Analysis in DC Electrical Power System

The CMA process is illustrated below by taking the "Total Loss of DC Network," one of the EPS failure conditions of a certain aircraft as an example. This case is simplified, and due to the differences in specific projects, the analysis and the results of this CMA case may not be applicable to other CMAs. Therefore, this case is for reference only.

6.2.5.1 Inputs

The inputs include the following information:
1. Independence principles and/or independence requirements identified by the PSSA
2. The CMA checklist
3. The EPS layout
4. The design considerations of protection against common causes
   Each item of the inputs is listed below, respectively.

Table 6.1 Independence requirements from PSSA (part of them)

REF	Safety requirements
24-ARC-1	The normal power system should not be affected by RAT emergency power system.
24-ARC-2	The normal power system is independent of the emergency power system.
24-DSR-3	The LIDG/RIDG is independent of the RAT.
24-DSR-4	The LIDG/RIDG is independent of the RAT manual release cable (considering the aspects of design, installation, specific risks, etc.).
24-DSR-5	The LIDG/RIDG is independent of the Weight-on-Wheel Signal interface with the landing system.
24-DSR-6	The LIDG/RIDG is independent of the avionics built-in testing signal.
24-DSR-7	The RGCU and RAT GCU are independent.

6.2.5.1.1 Independence principles and/or independence requirements identified by the PSSA

The following independence requirements, which are generated in the PSSA process of the EPS, are shown in Table 6.1.

From the architecture design, the normal power channel should be totally different from RAT channel, and they should be independent and dissimilar. By further carding, some of the independence requirements in Table 6.1 is equal to "the normal power channel should be independent and dissimilar with RAT channel." Therefore, the following CMA activities of this case are for the independence requirements of "the normal power channel should be independent and dissimilar with RAT channel."

6.2.5.1.2 The Common Mode Analysis Checklist

According to the generic checklist, analysts will incorporate the engineering experience to determine the CMA checklist. The related contents are listed in Table 6.2.

6.2.5.1.3 The Electrical Power System Layout

The details of the EPS layout described here can also support the PRA and ZSA. The layout will not be repeated in the PRA and ZSA sections.

The EPS consists of three isolated power supply channels: the left channel, the right channel, and the emergency channel.

The left AC channel consists of the LIDG (installed in the left nacelle), the LGCU (installed in the left front equipment rack of the E/E

Table 6.2 Example of an EPS CMA Checklist

Common cause type	Common cause sources	Common cause failures/errors
Design	The same software	The software common cause failure of LCGU, RGCU, and AGCU
	GCU	The failure of multiple power supply channels caused by a single control unit
	Layout and installation of equipment and wires	The common cause failure of left/right/emergency AC channels The common cause failure of left/right DC channels and ETRU power supply channels
	Common connector	The failure of multiple power supply channels caused by that of a single connector
	Common grounding point	The failure of multiple power supply channels caused by that of a single grounding point
Manufacturing	The same components processing and/or assembly procedures	The possible common cause failure caused by the similarity of components processing and/or assembly procedures
	Defective components	Defective components are installed in multiple LRUs of the same aircraft
	Electric components	The failure of multiple power supply channels caused by internal or external short
Installation	Installation error	The failure of multiple power supply channels caused by installation errors
Environment	EMI/HIRF/Lightning effects	The failure of power system equipment caused by EMIs, HIRF or lightning effects, such as GCU, RAT GCU, ADCU, IDG, AUX GEN, RAT GEN, TRU, and storage battery
	High temperature/fire	The failure of two main channels and emergency channel caused by high temperature or fire
	Environmental stress	The failure of multiple power supply channels caused by environmental stress

(Continued)

Table 6.2 (Continued)

Common cause type	Common cause sources	Common cause failures/errors
Maintenance	Maintenance procedure	The failure of multiple power supply channels caused by a maintenance procedure error
	Maintenance personnel	The failure of multiple power supply channels caused by maintenance procedure errors for operation personnel lacking skills
Operation	Operation procedure	The failure of multiple supply channels caused by operation procedure errors
	Operation personnel	The failure of multiple supply channels caused by maintenance procedure errors for operation personnel lacking skills

compartment), the Left Power Distribution Assembly (LPDA, installed under the power center), and the L AC BUS (installed on X1 plate of the power center), all of which are arranged in the left side of the aircraft. The main feeders of the L AC channel are laid along the left side of the aircraft, all the way from the LIDG, through the left pylon, the cargo compartment, and the EE compartment then the LPDA to the L AC BUS. The control wiring of the L AC channel is also laid along the left side of the aircraft, all the way from the LIDG to the LGCU through the left pylon, the cabin ceiling, and the left equipment rack of EE compartment.

The right AC channel consists of the RIDG (installed in the right nacelle), the RGCU (installed in the right front equipment rack of E/E compartment), the LPDA (installed in the station 120 of E/E compartment), and the R AC BUS (installed on X2 plate of the right power center), all of which are arranged in the right side of the aircraft. The main feeders of the R AC channel are laid along the right side of the aircraft, all the way from the RIDG to the Right Power Distribution Assembly (RPDA) and then to the R AC BUS. The control wiring of the R AC channel are also laid along the right side of the aircraft from the RGCU to the RIDG.

The emergency AC channel consists of the RAT generator (installed in the RAT compartment), the RAT GCU (installed in the forward accessory compartment), the RGLC (installed in the E/E compartment), and the AC ESS BUS (installed on X3 plate of the power center), with the equipment layout independent of the left and right AC channels. The main feeders of the emergency AC channel are laid from the RAT generator to the emergency power control box located in the E/E compartment and then to the AC ESS BUS on the X3 plate of the power center. Its control wiring are laid from the RAT GCU to the RAT generator. Either the laying of feeders or the control wiring of the emergency AC channel are independent from that of the left and right AC channels.

The left DC channel consists of the LTRU (installed in the forward cargo compartment), the LTRUC (installed in the power center), the LESSC (installed in the power center), the L DC BUS (installed on X1 plate of the power center), and the L DC ESS BUS (installed on X1 plate of the power center and left side of X4 plate on the roof). The main feeders of the L DC channel are laid along the left side of the aircraft from the LTRU to the LTRUC, which is in the power center. The left DC channel is mainly controlled and protected by the LGCU, with the control wiring laid along the left side of the aircraft.

The right DC channel consists of the RTRU (installed in the forward accessory compartment), the RTRUC (installed in the right power control box of the E/E compartment), the RDC BUS (installed on X2 plate of the right power center and the X3 plate of the power center), and the R DC ESS BUS (installed on X2 plate of the right power center, the X5 plate of E/E compartment, and the right side of X4 plate on the roof). The main feeders of the L DC channel are laid along the right side of the aircraft from the RTRU to the RTRUC, which is in the E/E compartment. The right DC channel is mainly controlled and protected by the RGCU with control wiring laid along the right side of the aircraft.

The emergency DC channel consists of the ETRU (installed in the forward cargo compartment), ETRUC (installed in the emergency power control box of E/E compartment), and DC ESS TRANSFER BUS (installed in the X3 plate of the power center). The feeders of the emergency DC channel are laid from the ETRU to the ETRUC in the E/E compartment, independent of the left/right DC channel. The emergency DC channel is controlled by the relay with different control modes from that of the left/right DC channel, and the laying of its control wiring is independent of the left/right DC channel.

6.2.5.1.4 The Design Precautions Against Common Cause Failures

Before CMA, some design precautions are adopted in the preliminary design architecture based on the previous experience of EPS. The design precautions will be updated after performing CMA, which is an iterative process.

1. Descriptions of the functional independence and isolation of the EPS
   a. The independence and isolation of the Bus
      i. The Buses are designed to be isolated. All of the tie relays are disconnected under normal operation, with LIDG and RIDG supplying power to the left and right channels, respectively. Therefore, any failure or error of the system affects only one channel, meanwhile, the control system can be designed to prevent failure from spreading to the possible zones.
      ii. The tie capability of the AC Bus. With one or two GCU failures, the software-controlled AGCU will control the LACTR and RACTR to provide AC power.
      iii. The tie capability of the DC Bus. With the failure of one channel, LGCU or RGCU controls DCTR to provide DC power (the redundant power supply of the left/right DC Bus).
      iv. The tie capability of DC ESS BUS; in the event of a failure, the LETR and RTER, respectively, controlled by the auxiliary contact logic provide tie of the DC ESS BUS, which can be supplied with time-limited power by storage batteries through the use of relay logic and drive-by-wire.
      v. In the case of system failure, the power can be supplied to the essential loads by removing part of the load.
   b. The independence and isolation of the power
      i. Each IDG is installed in its own corresponding engine, from which the mechanical energy is extracted.
      ii. The failure of a single IDG does not cause that of another IDG.

      Therefore, this mode of the power supply and that taking IDG as the power are mutually independent and isolated. Although the RAT system is an independent redundant power, not all 100% of the components of the RAT system are detectable. In the case of emergency, the loss of the RAT system is considered an unsafe condition.

    c. The independence among GCUs

       i. Due to the independence and isolation from each other, the loss of any GCU will not affect the function of the remaining GCUs.

      ii. Each GCU has multiple power supply inputs, and each generator contains the permanent magnet generator to provide independent power for its GCU.

     iii. In the case of bus failure, the GCU can disconnect the related tie relay to prevent the failure propagation.

     iv. The fail-safe mode is established to prevent system actions or cascading failures.

      v. The fault detection and power-on self-test include power quality, software failure, overcurrent, controller fail-safe, loss of drive of the contactor, and overvoltage.

    d. Tie function

        To ensure the adequate safety of the EPS, the bus with power loss can be powered through tie function in the case of component or bus failure in the left and right channels. If a single IDG fails, the crew can start the APU generator and use the APU generator instead of the failed IDG to supply power to the corresponding bus through tie function.

2. Descriptions of the independence of the system equipment and harness layout

    a. The isolation of the main channel from the emergency channel

        Two IDGs and APU generators are installed within their own separate compartments and driven by independent generators. IDG is installed in the left/right nacelle; the APU generator is installed in the APU compartment; see Table 6.3 for the installation position of the main channel's contactor, relay, and fuse. The RAT generator is installed in the RAT compartment and releases RAT in the case of emergency. The ACETR, RGLC, and ETRUC are installed in the E/E compartment, and the ETRU is installed in the forward cargo compartment. The feeders from the L AC BUS or R AC BUS to the ACETR AC and that from the RAT generator to the RGLC should be laid separately.

    b. The isolation of the left channel from the right channel

       i. The isolation of the left channel equipment from the right channel equipment: see Table 6.3 for the left/right channel equipment layout. As shown in Table 6.3, the left channel

Table 6.3 The equipment layout of left and right channel

Left channel		Right channel	
Equipment title	Installation location	Equipment title	Installation location
LIDG	Left nacelle	RIDG	Right nacelle
LGCU	left front	RGCU	Right nacelle
Left Power Distribution Assembly (LPDA)	Left power center	RPDA	E/E compartment
LTRU	Left power center	RTRU	Forward accessory compartment (left)
LTRU Contactor	Left power center	RTRU Contactor	Left power supply box
LESS Contactor	Left power center	RESS Contactor	Left power supply box
LESS Tie Relay	Left power center	RESS Tie Relay	Left power supply box
LDC Tie Fuse	Left power center	RDC Tie Fuse	Left power supply box
LESS Fuse	Left power center	RESS Fuse	Left power supply box

equipment is concentrated on the left side of the aircraft, and the right channel equipment is concentrated on the right side.

ii. The isolation of the left channel feeder from the right one: the independent and isolated laying is required for the cables of the left/right AC and DC systems. Feeders from the LIDG to the LPDA belong to the left AC feeder harness, from the RIDG to the RPDA belong to the right AC feed harness, from the LTRU to the LTRU C belong to the LTRU DC feeder harness, and from the RTRU to RTRU C belong to the RTRU DC feeder harness. Therefore, the independent binding of AC and DC feeders is required to achieve the physical isolation. The L AC BUS and the L DC BUS are mainly concentrated on the circuit breaker plate of the power center and the top control panel. The R AC BUS and the R DC BUS are distributed on the circuit breaker plate of the right power center and under the power center behind the right console. The power wiring of the LDC ESS BUS and RDC ESS BUS, which interconnected to the DC ESS Transfer BUS, are physically isolated. The power wiring from the LDC ESS BUS via the LETR and LTF to the DC ESS Transfer BUS power is single wire harness and distributed in the power center. The power wiring from the RDC ESS BUS via the RETR and RTF to the DC ESS Transfer BUS belongs to the interconnected and single left/right DC emergency wire harness, which are distributed in the EE compartment. In summary, the wiring layout basically ensures the physical isolation of the left channel from the right channel in the power supply system.

iii. The AC ESS BUS and DC ESS BUS feeders should be physically isolated: under normal conditions, the AC ESS BUS is supplied by the main AC power through the ACETR and RGLC. Under emergent conditions, the emergency generator is powered by the RGLC, which belongs to the AC ESS BUS harness in the E/E compartment. The LDC ESS BUS feeders are concentrated on the power center and distributed on the left side of the aircraft. The RDC ESS BUS feeders are concentrated on the right power box, which are distributed on the right side of the aircraft. The wiring from the ETRU through the ETRU C to the DC ESS Transfer BUS is distributed between the forward cargo compartment and the EE compartment. Thus, the AC ESS BUS and DC ESS BUS basically realize effective physical isolation through the wiring.

iv. The isolation of the left channel control wiring from the right control wiring: the LGCU is placed on the left equipment rack of the E/E compartment and, via its left side to the left power center, achieves control of the LPDA, LTRUC, LESSC, LETR, and other equipment. The RGCU is placed on the right front rack of the E/E compartment, allowing direct control to the R PDA, RTRUC, RESSC, and RETR in the right equipment rack.

3. Descriptions to specification/operation/maintenance/test/calibration integrity

Specifications applied in the power system design and manufacturing are generally recognized by the international aviation industry and civil aviation authorities, which can guarantee the integrity.

The operation/maintenance/test/calibration procedures for the EPS are reviewed many times by the company and the certification authority to ensure the integrity and avoid common cause failures.

It is required to conduct intensive and comprehensive training and assessment for the operation/maintenance/test/calibration personnel and to establish rules and regulations implemented according to procedures to avoid common cause failures.

6.2.5.2 Common Mode Analysis process

According to the PSSA, there is one "AND" gate under the "Total Loss of DC Network," and the three events under this gate are the "Loss of LDC BUS power supply", "Loss of DC ESS Transfer," and "Loss of R DC BUS power supply". The independence between the normal and emergence power system must be confirmed. To illustrate the main process of CMA, the following CMA activities are performed for the independence requirements of "the normal power channel should be independent and dissimilar with RAT channel" in Table 6.4.

6.3 PARTICULAR RISK ANALYSIS

6.3.1 Introduction

The objective of PRA is to analyze typical events in the history of civil aviation that affect the independence among each channel of the redundancy system to ensure the actual occurrence of the redundant function.

Some of these risks might conform to specific airworthiness requirements (e.g., uncontained engine rotor failure and tire burst), while others are subjected to the known threats external to the aircraft and systems. Typical risks include but are not limited to fire, uncontained engine or APU rotor failure, bird strike, tire burst, ice, hail and snow, wheel rim release, lightning, HIRF, and hydraulic accumulator burst.

Other specific risks are from high-pressure storage bottles, high-pressure duct rupture, high-temperature duct leakage, and leaking fluids (e.g., fuel oil, hydraulic oil, acid battery, and water). Despite of these risks normally being analyzed within the scope of ZSA, additional analysis is occasionally needed.

Each risk has been determined and regarded as a specific research subject. This is for the purpose of ensuring that any safety effects are eliminated or that their risks are acceptable.

Table 6.4 CMA worksheets

Common cause type	Common cause sources	Possible common cause failure or error	Common cause failure or error analysis	Conclusion
Design	The same LRU software	Common cause failure of software in the LGCU, RGCU, and AGCU	The software-contained units, which supply power to the DC bus, are the LGCU, RGCU, and AGCU. Among which, the software contained in the LGCU and RGCU are similar, so there may be common causes. However, to prevent total loss of DC power, the software contained in the RAT GCU is different from that in the LGCU and RGCU.	No common cause failures or errors are caused by the same LRU software.
	Common controller	Failure of multiple supply channels caused by that of a single controller	The LGCU controls the left power supply channel (including the LIDG and the associated contactors in the left power supply channel) while providing partial the redundant control for the remaining channels. The RGCU controls the right power supply channel (including the RIDG and associated contactors in the right power supply channel) while providing partial redundancy control for the remaining channels. The AGCU is responsible for the control of the auxiliary supply channels (including the control of the APU GEN and the AGR) and provides partial redundancy control for the remaining channels; the RAT GCU is responsible for the control of the RAT GEN and RGLC. Therefore, the failure of either controller does not result in the failure of multiple power supply channels.	No multiple power supply failure will be caused by single controller failure.

The installation of equipment and wiring harness	The common cause failure in the left and right DC channel and ETRU power supply channel	According to the EPS installation layout, the installations of the equipment and wiring harness of the left/right DC channel and the ETRU power supply channel are independent of each other, and there is no common cause failure.	No common cause failure is caused by the installation of the equipment and wiring harnesses.
Common connector	Failure of multiple power supply channels caused by that of a single connector	Based on the three DC power supply channels, the other connectors are not shared by the three channels, except for the connectors R5/P1-2420 and the connectors R5/P1-2430. If the connector R5/P1-2420 fails, it may lead to the LTRUC and DCTR open; at this point, only L DC BUS failure will not lead to multiple power supply channel failure simultaneously. If the connector R5/P1-2430 fails, it may lead to the RESSC, ETRUC, DCTR, LETR, RETR unable to close, and the R DC ESS BUS and DC ESS TRANSFER BUS will fail to supply power at this point. However, the L DC BUS, R DC BUS, and L DC ESS BUS are still valid, so the classification of this failure condition is Major. In conclusion, common connector failure will not cause the failure of more than one DC power supply channel, and the installation of different connectors of DC power supply channels in the EPS can avoid common cause failure.	No multiple power supply failure will be caused by single controller failure.

(Continued)

Table 6.4 (Continued)

Common cause type	Common cause sources	Possible common cause failure or error	Common cause failure or error analysis	Conclusion
	Common ground point	Failure of multiple power supply channels caused by that of a single ground point	The grounding stud in the left, right, and emergency DC channel is divided into the main power grounding stud and signal circuit grounding stud. Through the design of strong and weak separation, the mutual interference is avoided when the power supply channel works properly. There are no common ground points except for the grounding stud G3–2403 (common to the left DC and emergency DC channel) of the three DC channels. The installation of grounding stud for different channels are independent of each other. G3–2403 in the emergency DC channel is mainly to provide the AGR and EPR contact state to detect circuit grounding, so its failure does not affect the emergency DC channel. As a result, multiple DC supply channels will not fail due to the failure of a single ground point.	Multiple power supply channels will not fail due to the failure of a single ground point
Installation	Installation error	Failure of multiple power channels caused by installation errors	Different equipment in the EPS taking a different connector pin will not lead to installation errors. The same equipment in the EPS installed in the adjacent location are the LGCU and AGCU. Because these two equipment are identical, the system functions are not affected, even if the plug on the cable is incorrectly plugged into another equipment. Therefore, the equipment installation errors will not cause the failure of multiple power supply channels.	Multiple power supply channels will not fail due to installation errors.

| Manufacturing | The same components processing and/or assembly procedures | Common cause failure caused by similarities in component processing and/or assembly processes | TRU, the main battery (including the APU battery) and part of DC equipment in the EPS are provided by different suppliers. Therefore, the main power supply channel and the emergency power supply channel of the EPS do not have the same components processing and/or assembly procedures and will not cause the failure of the main power supply channel and the emergency power supply channel simultaneously. | Meet the safety requirements. |
| | Defective components | Defective components are installed in multiple LRUs of the same aircraft | TRU, the main battery (including the APU battery) and part of DC equipment in the EPS are provided by different suppliers. TRU are totally different from batteries. The LTRU and RTRU are controlled by the corresponding GCU which based on the digital signal processor. The ETRU is controlled by the relay under the charge of the crew, and the main battery/APU battery is controlled by the bus controller which based on the FPGA. Therefore, neither the TRU nor the main battery/APU battery control unit exists identical parts. Based on the above analyses, there are no identical components between the equipment of the main channel and the emergency channel. Therefore, even if defective components are installed in the multiple LRUs of the EPS of the same aircraft, the LRU failure due to these defects cannot occur simultaneously, thus not causing common cause failure. | Meet the safety requirements. |

(Continued)

Table 6.4 (Continued)

Common cause type	Common cause sources	Possible common cause failure or error	Common cause failure or error analysis	Conclusion
	Electric components	Failure of multiple supply channels caused by internal or external short	The installation of equipment and wiring harness of the left, right and emergency power supply channels are independent of each other. The DCTR, LETR, and RETR contactors are in the open position during normal operation of the EPS, so the three channels are isolated from each other and operate independently. The EPS is equipped with circuit breakers, fuses and contactors to protect against possible shorts. As a result, the failure of multiple power-supply channels will not be caused by internal or external shorts.	Meet the safety requirements.
Environment	EMI/HIRF/ lightning effect	Failure of EPS equipment caused by EMI, HIRF or lightning effects	The protection of EMI/HIRF/Lightning for the EPS equipment is defined in terms of the safety objectives defined in the EPS SFHA, in conjunction with the operating environment of the EPS equipment. Equipment qualification tests show that the EPS equipment can meet the defined EMI/HIRF/Lightning requirements. Therefore, according to the safety objectives of EPS, the protection requirements for EMI/ HIRF/Lightning of power system equipment are developed through the design process to limit the occurrence probability of common cause failure to meet the safety requirements of the EPS.	Meet the safety requirements.

	Environmental stress	Failure of multiple power supply channels caused by environmental stresses	According to the installation location and operation environment of the EPS equipment, combined with DO-160D, the environmental stress requirements of the EPS is defined in General System Requirements Document of the EPS, including vibration, temperature, altitude, humidity, etc. The EPS equipment qualification tests show that the EPS equipment can meet the environmental requirements. Therefore, the EPS equipment for the design of environmental stress can meet the relevant safety requirements.	Meet the safety requirements.
	High temperature/fire	Two main channels and emergency channel failure caused by high temperature or fire	The installation of equipment and wiring harness of the left, right and emergency AC channels are independent of each other, and the installation of equipment and wiring harness of the left/right DC channel and the ETRU power supply channel are independent of each other. Therefore, high temperature/fire will not lead to both failure of the main channel and emergency channel.	Meet the safety requirements.
Maintenance	Maintenance procedures	Failure of multiple power supply channels caused by maintenance procedure errors	The maintenance procedures are reviewed by the company and the certification authority to ensure the correctness and completeness.	Meet the safety requirements.

(*Continued*)

Table 6.4 (Continued)

Common cause type	Common cause sources	Possible common cause failure or error	Common cause failure or error analysis	Conclusion
	Maintenance personnel	Failure of multiple power supply channels caused by maintenance staff lacking of skills and maintenance errors	The maintenance personnel are required to receive a strict training and review.	Meet the safety requirements.
Operation	Operation procedures	Failure of multiple supply channels caused by operating procedure errors	The operation procedures are reviewed by the company and the certification authority to ensure the correctness and completeness.	Meet the safety requirements.
	Operation personnel	Failure of multiple power supply channels caused by the maintenance staff lacking of skills and maintenance errors	The operation personnel are required to receive a strict training and review.	Meet the safety requirements.

Documenting the review of the effects of each particular risk includes: the explanation of the particular risks being analyzed, the equipment affected by the particular risks and their installation zones, the failure modes result from the particular risks, and their failure condition effects and classifications on aircraft. Cross-checks to the effects on aircraft and the PSSA/SSA concerned shall be conducted. In addition, the consistence of "effects on aircraft" and "the classification" in PRA and PSSA/SSA shall be confirmed. The solution to any deviation from the initial assumptions and any problems emphasized after analysis shall also be included in this document. The certification authority shall focus on any problems that are uncovered during PRA.

The process of PRA is to conduct analysis on each risk one by one. The main PRA process is as follows:

- Define the nature of the particular risks to be analyzed. (e.g., tire/wheel burst)
- Define the failure model to be used for the analysis (e.g., the engine rotor burst model and tire burst model)
- List the requirements that shall be met (e.g., §25.903(d)(1) and §25.729(f))
- Define the affected zones/areas (e.g., landing gear bays)
- Define the affected systems/equipment and cross-check with ZSA
- Identify the accetped prevention measures for design and installation and perform cross-checks with the design and installation guidelines used in ZSA
- Review the consequences of particular risks on the affected equipment by cross-checking with FMEA/PSSA
- Assess the effects of particular risks on the systems/equipment of the aircraft by cross-check of the failure effects in the SSAs. The consistence of "effects on aircraft" and "the classification" in PRA and SSA shall be confirmed
- Determine if the consequences are acceptable. If acceptable, prepare justifications, and if not, prepare a design modification.

PRA should be carried out during the overall aircraft development process of a new aircraft or any major modification to an existing aircraft. Initially, drawings or models should be used. As the project progresses, the analysis shall be carried out based on mock-ups and then the completed aircraft.

6.3.2 Case Study

Many specific risks can affect the independence among the channels of EPS, such as uncontained engine rotor failures, tyre burst, and bird strike. Here, only the analysis of uncontained engine turbine rotor failure of EPS has been introduced to show the objectives and process of PRA.

The case is an analysis process of the effects of an engine uncontained turbine rotor failure on the safety of the EPS. The analysis methods and process have referred to FAA AC 20-128A [2]. The detailed information of layout and design precautions of EPS have been included in the case study of CMA.

1. The design considerations and precautions of the EPS
   a. Design considerations

   To minimize the harm to the EPS caused by uncontained turbine rotor failure, a certain type of EPS adopts the following design considerations:
   i. Placing the essential equipment and wiring harnesses outside the possible fragment impact zones
   ii. If within the fragment impact zones, making a backup for essential components or system and separating and isolating the system to ensure the efficiency of the redundancy system.
   b. Design precautions

   The EPS is composed of two main power channels and an emergency channel.
2. The layout of the EPS

   The EPS consists of three mutually isolated power channels: the left channel, right channel, and emergency channel.
3. Analytical assumptions
   a. Assuming that the energy of rotor fragments is infinite to destroy all pipelines, cables, and unprotected structures it passes, and this will not change the direction from its original track unless there is a deflection shields. However, protective shielding or an engine being hit could be considered strong enough to withstand the most energy of the fragments.
   b. Within the maximum range of the fragment spread angle, the spread probability of rotor fragments in all directions is equally distributed.
   c. The components of the system are deemed a complete failure after its outer profile is hit by fragments.

d. The analysis shall address the effect of a single rotor burst on the system during a single flight; its probability is 1.0, assuming that there is only one piece of fragment acting on the aircraft when considering the effect of one-third disk fragments on aircraft.

e. The rotors of two engines will not burst simultaneously

This case does not involve analysis of the probability of equipment failure after being hit by magnitude fragments; for this reason it does not involve calculation of the entry angle and exit angle of the rotor fragments when the engine rotor is burst.

4. Failure conditions and safety analysis of the EPS caused by rotor burst

Analyze the influence zones of the engine rotor fragments by defining the uncontained engine rotor burst failure modes and establishing the uncontained engine rotor burst failure model. According to the analysis results, the influence zones of rotor fragments are mainly distributed in the rear cargo bay, cabin, engines, pylon, recirculating-fan compartment, and rear attachment compartment. From the perspective of aircraft safety, when considering uncontained rotor burst, only consider the system or equipment that causes damage to the aircraft flight safety. On this basis, and combined with the layout of the EPS, it can be concluded that the equipment and wiring harness of the EPS being affected in this area include the LIDG and its feeders and control cables, RIDG and its feeders and control cables, APU GEN feeders and control cables, APU batteries, control units, control boxes and related wiring harness.

Taking the rotor burst of the left engine as an example, the analysis is as follows, which is similar to the rotor burst of the right engine. See Table 6.5 for worksheets of analysis of "Left engine rotor burst."

5. Summary

The analysis of the impact angle of the EPS equipment and wiring harness hit by rotor fragments along with the qualitative analysis of the impact on the EPS indicate that the most severe effect of the EPS is a hazardous failure condition in the case of uncontained rotor burst of the engine, i.e., "Loss of the main AC power" (L AC bus failure and RAC bus failure).

As the emergency power supply system and the system affected by the rotor burst are completely independent and fully isolated, the aircraft can continue its safe flight and landing with the essential electrical load equipment of the aircraft supplied by the emergency power

Table 6.5 Worksheets of analysis of "Left engine rotor burst"

Events	The affected equipment	Effects	The related failure conditions	Classification	The safety analysis
Left engine rotor burst	LIDG and its feeders and control cables, APU GEN feeders or control wiring harness	The LIDG fails due to loss of drive, and rotor fragments hit the APU GEN feeder harness or control wiring harness	Loss of left main AC power and APU power	Major	There is still a main generator to supply power. There is a reduction in functional capabilities or safety margins. Kitchen and other nonessential loads are automatically unloaded to ensure important and essential electrical load equipment are supplied without affecting the safe flight and landing.
	LIDG and its feeders and control cables, RIDG and its feeder and control cables, APU GEN feeder or control wiring harness	The LIDG fails due to loss of drive, and the right AC channel (RIDG, RIDG feeders, or RIDG control harness) and the APU channel (APU GEN feeders or control harness) are hit by rotor fragments	Loss of left and right main AC power and APU power	Hazardous	Part of the AC electrical loads cannot work. The system gets into an emergency condition, and the emergency power (RAT GEN, the main battery and APU batteries) supply for the loads. There is a reduction in functional capabilities or safety margins. When the system get into emergency condition, the power supply will be interrupted and then be resumed after the interruption.

| LIDG and its feeders and control cables, RIDG and its feeders and control cables, APU GEN feeders or control harnesses, APU batteries, control units, control boxes, and related wiring harnesses | LIDG fails due to loss of drive, and the right AC channel (RIDG, RIDG feeder, or RIDG control harness), the auxiliary AC channel (APU GEN feeder harness or control harness), and the APU battery are hit by the rotor fragments | Loss of left and right main AC power, APU power and APU batteries | Hazardous | Part of the AC electrical load cannot work. The system gets into the emergency condition, and the emergency power (RAT GEN and the main battery) supply for the loads. There is a reduction in functional capabilities or safety margins. When the system enters the emergency condition, the power supply will be interrupted and then be resumed after the interruption. |

Note: The effects of the feeder harness (any one, two or three phases) and the control wiring harness being hit by rotor fragments are the same, which is the loss of power to the ipsilateral channel.

after the engine rotor burst. Thus, the safety requirements of the EPS can be satisfied.

6.4 ZONAL SAFETY ANALYSIS

The purpose of the ZSA is to ensure that the system under analysis is designed and installed to meet the safety requirements related to the following four aspects.

1. The basic guidelines for the design and installation
2. The effects of failures on aircraft
3. Maintenance errors
4. The independence declared by FTA "AND" events.

Therefore, the installation of systems and equipment shall meet the requirements as follows:

- *Basic Installation Requirements*: Check that the installation design on the aircraft is appropriate and meets the installation requirements.
- *Interface Requirements Between Systems*: The effects of equipment failures on other systems and structure should not be beyond the range of their physical activities.
- *Requirement of Maintenance Error*: Consider the effects of installation and maintenance errors on the system or aircraft.
- Using isolation technology to ensure that the design and installation of redundant systems meet the requirements of independence.

By partitioning aircraft into several zones, the ZSA evaluates the physical installation of equipment and systems to identify potential hazards caused by mutual influences between equipment installed on the aircraft as well as the influences of the operation environment on such installed equipment. When the influences are likely to be related to safety, ZSA should identify and analyze the influences, then change the design or show in the appropriate safety assessment activities that it is acceptable.

First, establish the design and installation guidelines for a new aircraft; this is to ensure the independence inside the zone. For existing aircraft, it can be assumed that the design and installation guidelines have been inspected and met previously. For derivative aircraft modifications, the guidelines of the baseline aircraft should be used as much as possible.

Second, inspect the installation in each zone of the aircraft and evaluate whether the guidelines have been met. The aircraft zones partitioned for maintenance purposes can be used for this inspection.

Finally, prepare a list of systems or equipment in each zone of the aircraft, which may be based on installation drawings, mock-ups, or aircraft (depending on the development process). For each system or equipment on the list, a list of intrinsic failure modes that can affect adjacent systems or equipment should be established. This failure modes list may be based on FMEAs or FMESs of the systems or equipment and the intrinsic hazards of the systems or equipment.

The FMEA should consider the failure modes of the system or equipment, the external effects, and the effects on the aircraft. The effects of these failure modes on the adjacent systems should be judged based on the system specification, PSSA, or equivalent activities. The described effects on the system/aircraft should be consistent with the relevant SSA or ASA. The ZSA should consider the external effects on the systems or equipment as a potential common cause failures in different systems. FTA is a tool that can be used in assisting the analyst in consideration of these potential common cause failures.

The ZSA report shall include the results of zonal inspections against the design and installation guidelines and the systems or equipment external failure effects on the aircraft. The relevant SSA should be referenced any acceptable results. Any deviation from the design guidelines should be considered a candidate for design modification.

Although initial ZSA inspections can be carried out using drawings or models during the development process, the final certification is preferably based on the conclusion of a physical inspection of the completed aircraft.

When conducting a zonal safety analysis, the following hazard sources shall be considered:

1. Devices with high-energy rotor, such as engines, air recycler, APU, fans, engine generator, and hydraulic pumps.
2. All components with potential high-energy release, such as pressure bottles, accumulators, oxygen cylinders, fire extinguishers, cold gas cylinders, tires.
3. Any equipment carrying corrosive substances that may cause deterioration of the surrounding environment, such as waste water pipelines, fuel tanks and ducts, batteries, and the equipment of the hydraulic system.
4. Any possible pipes and pipelines that might cause high-temperature and high-pressure gas leaks, such as the engine exhaust ducts, high-temperature pipelines of the environment condition system, and anti-ice ducts.

5. All electrical equipment and wires that are likely to cause smoke or fire after failure or overheating.
6. Equipment and components that are in movement when normally operated, such as flight control devices and landing gear control devices.

Partitioning aircraft into zones, identifying the hazard sources and the list of systems/equipment in each zone shall be jointly completed by the safety engineers and the general departments. The system safety analysis in each zone should be completed by the safety engineer. The zonal analysis results of the system shall be an integral part in the SSA report.

In fact, there are some overlaps between CMA and ZSA activities, such as the check of physical separation of different channels of EPS.

A simple example of a zonal safety analysis for EPS is as follows:

Zones: Forward cargo compartment.

Hazard Sources: Waste water discharge line.

The affected equipment in the EPS: ETRU, LTRU.

Failure modes: waste water discharge pipeline leakage, waste water condensed droplets, corrosion of equipment, and the device abnormal output

The effects of failure mode on EPS: The ETRU cannot supply power to the DC Essential Transfer Bus, and the LTRU cannot supply power to the left DC Bus.

Protective precautions: Change the waste water line connector above the TRU to the ceiling of the forward cargo compartment.

REFERENCES

[1] [SAE96] SAE ARP 4761. Guidelines and methods for conducting the safety assessment process on civil airborne systems and equipment. Society Automotive Engineers; 1996.
[2] AC20-128A. Design considerations for minimizing hazards caused by uncontained turbine engine and auxiliary power unit rotor failure. FAA; 1997.

CHAPTER 7

Failure Modes and Effects Analysis with Summary

Contents

7.1 CONCEPT

In the early 1950s, Failure Modes and Effects Analysis (FMEA) was first adopted by the Northrop Grumman Corporation in the development of an aircraft primary control system, gained a good result. In the early 1960s, FMEA was used by NASA for Project Apollo. Since then, FMEA has been applied to the design and development of military systems, such as aviation, aerospace, vessel, and weapons. In the 1970s, MIL-STD-1629, *Military Standard Procedures for Performing a Failure Mode, Effects, and Criticality Analysis*, was issued, in which the principle of conducting FMEA was clearly specified. In 1984 the Unite States issued an

Civil Aircraft Electrical Power System Safety Assessment
DOI: http://dx.doi.org/10.1016/B978-0-08-100721-1.00007-8

update version of this standard, i.e., MIL-STD-1629A/notice 2. Currently, FMEA has been widely applied to the safety and reliability assessment process.

FMEA is usually used to get consideration of all possible failures and their effects on operation; to show which criteria must be given special consideration in development and tests; to provide a listing of all potential failures arranged according to the importance of their effects; to provide basis documentation necessary for further analyses.

In SAE ARP4761, FMEA is described to be one of the safety analysis methods [1]. FMEA is an analysis method intended to identify the failure modes of a system, item, function, or piece–part and determine the effects on the respective level as well as higher level of the design. According to the needs of safety analysis, FMEA may be qualitative or quantitative, applicable to any type of system. Besides all items in qualitative FMEA, quantitative FMEA also calculates the failure rate of each failure mode from methods introduced in Section 7.4. The results of FMEA may support other analysis (such as FTA and DD) used in SSA.

FMES is an extract from the respective FMEA(s) by summarizing lower level failure modes that lead to the same effects. Summaries of these failure modes are needed as an input to FTA basic events.

7.2 FAILURE MODE AND EFFECTS ANALYSIS

To meet the need from SSA, there are two basic FMEA types can be chosen, namely functional FMEA and piece–part FMEA. The main difference is that functional FMEA considers the failure modes of functions, while piece–part FMEA considers the failure modes of components. Mostly, functional FMEA is the first choice, as less design details within subsystem are considered than piece–part FMEA. There are three situations when piece–part FMEA may be done: the more conservative failure rate determined by functional FMEA cannot satisfy FTA failure probability budget; for systems rely on redundancy; or for mechanical items and assemblies. For most time, functional FMEA and piece–part FMEA are similar. So we will not distinguish them in the rest of this section.

7.2.1 Inputs

Before starting FMEA, it is usually required to obtain the following information:

1. FMEA requirements from other safety analysis tasks
2. Specifications in analysis level
3. Current drawing
4. Parts list of every system or item
5. Functional block diagram
6. Explanatory materials, including operating principles
7. Available list of failure rates (Section 7.4)
8. FMEAs of previous or similar products
9. The available preliminary list of component failure modes formed by previous FMEA (if available).

7.2.2 Process

To performing FMEA, it should be noted that organizing a team from multiple disciplines is quite useful. It will help to assure multiple viewpoints and can sharpen the focus of the results to account for various possible sources and interaction of issues and their cause. The design engineer knows more than anyone how a system or equipment works: what is important to its operation and what is not. FMEA is best performed with a team including design engineer, test engineer, maintenance engineer, safety engineer, and reliability engineer.

When starting FMEA, it is important to determine the breakdown level. FMEA is usually performed in component level, but it can be used in subsystem level or device level. Then the effect of each failure mode is determined at the given level.

All safety-related effects and any other effects identified by the requirements, should be considered in FMEA. In cases where it is not possible to identify the specific nature of a failure mode, the worst case effect should be assumed. If the worst case is unacceptable for the FTA, the failure modes must be examined at the next lower level.

In this chapter, we designed a FMEA worksheet (Table 7.1) considering the failure data for safety, the identification and corrective actions for maintenance and testability, as well as the dispatch requirement for airline service. Besides, by filling this worksheet step by step, all FMEA objectives in SAE ARP4761 can be accomplished.

Table 7.1 FMEA worksheet sample

FMEA#	Failure mode/ cause	Flight phase	Failure effect	Identification and corrective action of failures	Dispatch requirements with failures	Component failure rate	Mode failure rate	Exposure time	Mode probability	Classification	Remarks

Column 1: FMEA#—indicates the sequence number of the product in the analysis.

Column 2: Failure mode/cause—This column contains two aspects: failure mode and failure cause.

All the failure modes of component or hardware shall be included when filling in the form, and all of the failure modes shall be listed separately. Failure modes of system can be generally obtained as follows:

- For a new component, analysis can be conducted according to functional principles or architecture features, as well as failure modes of similar components.
- For components provided by suppliers, failure modes should be obtained from suppliers or analyzed and judged based on those that have occurred in components with similar functions and structures.
- For existing components, new failure modes can be determined by referring to those that have occurred in components during previous operation, along with analyses and corrections according to the similarities and differences of the operating environment.
- For common components and parts, their failure modes can be determined by various standards and handbooks
- For components with multiple functions, all possible failure modes of each function should be identified.
- In different phases the failure modes may be different.

Failure causes explains why the failure mode occurs. Failure causes should be analyzed from two aspects: first, the direct reasons causing failures or potential failures, such as the process of physical and chemical changes in the component; second, the reasons for failures generated by external factors (such as other system failures, equipment for measuring and testing, utilization, environment, and human factors).

Column 3: Flight phase—as defined in FHA.

Column 4: Failure effect—This column considers the effects of failure modes in column 2 of the component or hardware being analyzed during the flight phase in column 3. The effects shall be considered from three levels: local effect, higher level effect, and final effect (e.g. effect on aircraft).

Column 5: Identification and corrective action of failures—usually content following items:

1. *indication to the flight crew* which demonstrates that whether there is an indication to the flight crew when the failure mode in column 2 occurs.

2. *other failures with the same indication*, which lists other failures that will give the same indication to the flight crew in the subsystems and other systems in which the product is being analyzed.

3. *identification, isolation, and corrective actions of failures taken by the flight crew.* This column shall indicates a clear identification, isolation procedures, and corrective action of the failure. If there is a failure identification method, the most direct procedure which allows the flight crew to identify failures shall be listed. The identification procedure requires a specific measure or series of measures followed by inspection or cross-referencing of instruments, operating controls, circuit breakers, etc., or a combination of them. If a failure identification procedure is followed, effective corrective action can be determined.

4. *the effects of the possible error corrective actions*, taking into account any possible error corrective action and the effect of the measure on the system.

5. *failure isolation*—maintenance personnel, this column shall be filled with the defined maintenance requirements for the failure mode in column 2.

6. *corrective action*—maintenance personnel, this column shall be filled with the maintenance measures to be taken after the occurrence of the failure mode in the column 2.

Column 6: Dispatch requirements with failures—usually include the following items:

1. *If able, the aircraft can be dispatched*, this column shall be filled with "Able" or "Disable." After analyzing the effect of the failure mode, whether the aircraft can continue to be dispatched or the system with failure components can continue to run shall be assessed.

2. *If able, what is the flight limitation,* this column shall be filled with clear flight limitation on dispatched aircraft with failure.

Column 7: Component failure rate. This column shall be filled in with the total failure rate of the component, including failure rates for all failure modes of the component. For failure rate, obtaining methods are discussed in Section 7.4.

Column 8: Mode failure rate—This column shall be filled in with the failure rate of the failure mode in column 2.

Column 9: Exposure time—The period of time between when an item was last known to be operating properly and when it will be known to be operating properly again. Usually the measurement of exposure time includes the following three methods:

- Measured by continuous flight duration: when inspecting the equipment at the beginning of the flight, the exposure time is a typical average continuous flight duration.
- Measured by the maintenance check intervals: when checking the equipment after a defined maintenance interval, the exposure time is the maintenance interval.

Column 10: Mode probability—product of the failure rate of the failure mode and the exposure time.

Column 11: Classification—The classification shall be determined according to the failure effects in column 3. The classifications are the same as in FHA.

Column 12: Remarks—This column is used as a special explanation.

7.3 FAILURE MODES AND EFFECTS SUMMARY

To adequately address the need for inputs to higher level FMEAs and/or FTAs, FMES is discussed in SAE ARP4761. The failure rate of each failure mode in FMES is a sum of the failure rates coming from the failure modes of the individual FMEA(s) [1]. Table 7.2 gives a sample of common items in FMEA worksheet.

Table 7.2 FMES worksheet sample

FMES#	Failure mode	Failure rate	Exposure time	Classification	FMEA#

Column 1: FMES#—indicates the sequence number of the summary in the analysis.

Column 2: Failure mode—All causal (basic) failures which produce the same effect on the subsystem or component, shown by FMEA(s), are listed in the column "Failure Mode." That means: All identical failure effects listed in the FMEA(s) are summarized to one failure mode in the FMES. When necessary it is also the starting point for a top-down failure analysis (e.g., FTA) to find out the elementary failures causing that failure mode by combination.

Column 3: Failure rate—indicates the sum of the failure rates coming from the failure modes of the individual FMEA(s).

Column 4: Exposure time—the same as in FMEA.

Column 5: Classification—the same as in FMEA.

Column 6: FMEA#—indicates corresponding items in FMEA.

7.4 METHODS OF OBTAINING FAILURE RATE IN FAILURE MODES AND EFFECTS ANALYSIS

The quantitative analysis of a fault tree in a safety assessment needs to provide the failure rate data of basic events, which mainly come from quantitative FMEA and FMES. Failure rates are crucial for performing quantitative FMEA. At present, the obtaining of failure rate data mainly relies on such as in-service data, international standards or handbook data sources, and reliability tests. This section will briefly introduce these methods, and their specific theories can be found in relevant books.

7.4.1 In-Service Data

In-service data are valuable as it reflects the situations of the actual environment and maintenance condition, which can represent the actual performance much better than test in lab. It is generally believed that the entire reiteration of use condition is impossible by test in the lab, though test also needs to be conducted.

The in-service data analysis generally utilizes life distribution analysis and statistics inference methods. Finding out the type of life distribution from statistical analysis of in-service data is an important approach to analyze the product life and failure, predict failure development, and study the failure mechanism and designating maintenance strategies. On the basis of collected data the use of mathematical statistical methodologies can obtain the life distribution, compared to the phenomenon and cause of failure, by which the rationality of the life distribution can be judged. After defining the life distribution of products, parameter estimation of reliability data for different products can be performed in accordance with basic principles of mathematical statistics. With respect to the relation between the life distribution and reliability parameters, we can estimate various parameters for reliability design and analysis.

7.4.2 Standard/Handbook Data Source

7.4.2.1 MIL-HDBK-217

In 1957, the Reliability Analysis Center of the US Department of Defense officially issued MIL-HDBK-217, *Reliability Prediction Handbook for Electronic Equipment*, which has been updated seven times within more than 40 years. The US army canceled updating of MIL-HDBK-217 in 1995.

The basic idea for MIL-HDBK-217 is to provide an assessment for future system reliability by using historical part failure rate data. The scope of these parts ranges from microcircuits and discrete semiconductors to passive parts, such as resistors capacitors. For reliability predictions, it provides two basic methods [2]: the part count method and part stress analysis.

The part count method is applicable to reliability predictions in the initial design phase. In this period, the category and amount of parts are roughly fixed, while the specific work stress is not clear. It is a method to predict unit and system reliability via the number of each type of part (e.g., resistors, capacitors, diodes, and transistors) and the quality grade parameters. The accuracy of part count is poor, and the calculation results are conservative.

The part stress method is applicable to reliability predictions in the detailed design phase of electronic products. In this phase, most of designs have been completed. Part lists and the condition of part stress can be obtained. The idea of this method is to revise the basic failure rate in accordance with the factors, such as quality grade, environment, electric stress, temperature stress, and architecture characteristic. Its outcome is more accurate compared to the part count method and is closer to the actual condition. Therefore, it is usually used in the detailed design phase to find the weakness of reliability for taking measures to improve the design.

There are the following problems in MIL-HDBK-217:

1. The data used in the model parameters of the prediction handbook are from the 1990s and have not been updated.

2. The prediction value of reliability in the prediction handbook was formulated based on the average level of industrial, not aiming at specific providers and specific part types. Different processing and materials and the varied quality control level of different manufacturers may obviously result in differences in the reliability level.

Although there are various problems, as aforementioned, regarding MIL–HDBK–217, the aviation industry currently still adopts MIL–HDBK–217 to obtain failure rate data for electronic equipment where in-service data is lacking.

7.4.2.2 PRISM & 217plus

PRISM is a reliability prediction analysis method issued by the Reliability Analysis Center subordinate to the US Air Force in March 2000.

PRISM has overcome MIL–HDBK–217's shortcomings, i.e., that it performs reliability prediction based only on the product design and environment. By considering other factors affecting reliability, its prediction outcome can reflect the field reliability level.

PRISM has also taken account of the following factors on the basis of the part failure rate [3]:

1. *Process Factor*: PRISM has taken account of effects of process factor on product reliability, adopting many correction factors to quantitatively express the failures caused by process factor and define these correction factors by using a process grading method.
2. *Historical Data*: Many products have adopted inherited design methods, which conduct a certain degree of design change on the basis of old product models. When some historical data are generated in some old products, PRISM has provided an evaluation method by means of using historical data to adjust reliability prediction outcomes.
3. *Test or Field Data*: When there are some tests or field data to support current system designs, PRISM provides a Bayesian Analysis method to further adjust reliability prediction outcomes based on these data.

Different from other reliability prediction models, the distinctive characteristics are as follows:

1. Involve various factors affecting product reliability
2. Adopt all available information to evaluate field reliability
3. Conduct clipping in accordance with custom failure rate information
4. Adopt assessment methods of the quantitative confidence interval.

The characteristics of the PRISM methodology contain two meanings: the first one is to calculate the basic failure rate of each part/unit/system. The second one is to perform correction to obtain working failures by means of the basic failure rate calculated by using the process grade method and the Bayesian Analysis method. PRISM has taken account of both failure rates of parts constituting systems and many factors affecting product reliability, which makes its prediction outcomes more accurate and closer to the actual value.

The process of predicting reliability based on the PRISM method can be divided into three steps: (1) first, analyze various parts of systems and their information to predict working failure rate data of all parts by using a part reliability prediction model in PRISM; (2) perform reliability prediction based on reliability prediction outcomes of various parts to obtain working failure rate data of all parts; (3) perform system reliability prediction in accordance with reliability prediction outcomes of various parts to obtain final field reliability prediction outcomes of the system.

The principle for 217 plus is same as that of PRISM, and PRISM has been updated to 217plus. Both methods have performed correction for the initial failure rate by using process factors, such as design, manufacturing, part quality, system management and so on, which overcome the major shortcomings of MIL-HDBK-217.

217plus includes two parts: part-level reliability prediction and system-level reliability prediction. First, perform assessment for the failure rate of parts by using a reliability assessment model and then make a summary to evaluate the failure rates of parts and systems.

Compared to PRISM, 217plus has not covered part reliability prediction, but it contains a part-level reliability prediction process of system reliability in PRISM.

7.4.2.3 FIDES

FIDES is a modern reliability prediction plan, and its current version is FIDES-2009A.

FIDES adopts rich failure data from aviation, militariy fields, and manufacturing enterprises. The purpose of its development lies in providing pragmatic reliability prediction of electronic products in particular under various strict external environmental conditions. Therefore, this method targets reliability predictions for electronic devices, electronics and electrical products, such as integrated circuits, semiconductor discrete devices, capacitors, elector-regulators, resistors, potentiometers, sensors, transformers, electric relays, printed circuit boards, connectors, and piezoelectric elements.

The FIDES method has taken account of both the failures caused by internal factors of parts, such as product processing and quality distribution, and the failures caused by external factors, such as norms, designs, manufacturing and integration of products, and even the influences from the product supply chain [4].

FIDES has mainly taken account of three factors affecting reliability, i.e., the technology, manufacturing process, and application of products, which covers the entire life cycle of products from design and service to maintenance. The technology not only indicates itself but also contains the techniques integrated into products. Manufacturing process means all status of products from manufacturing and designs to being replaced. Application has taken account of product design, operation, and final use limitation for users.

7.4.2.4 Nonelectronic Parts Reliability Data and Electronic Parts Reliability Data

With extensive use of manual predication methods, such as MIL-HDBK-217, consumers have realized MIL-HDBK-217 can only provide prediction for electronic parts. Nonelectronic parts also need failure rates.

Thus, NPRD (nonelectronic parts reliability data) has emerged as a requirement that describes various original failure rate data of electronics, mechanical and electronic products, and mechanical parts and components. NPRD has mainly addressed two problems: (1) the provision of failure rates for various components that have not obtained failure rates through parts; (2) MIL-HDBK-217 or other prediction methods only contain the failure rate models of some electronic parts, which thus are complemented by NPRD to provide partial failure rate data not included in these models, such as electromechanical parts and mechanical parts.

The EPRD database and NPRD database can be complementary to each other, i.e., the data in these two databases are not duplicated. Therefore, they both can provide reliable calculation support for many categories of equipment and parts. The EPRD database covers the failure rate data of various electronic parts, such as capacitors, diodes, integrated circuits, optoelectronic equipment, resistors, thyristors, transformers, and transistors, and its current version is EPRD-2014. The NPRD database covers broader and more extensive failure rate data, such as electric equipment, electric–mechanical parts, and pure mechanical parts, and its latest version is NPRD-2016. Currently, NPRD is the main source for obtaining failure data for nonelectronic parts.

7.4.3 Reliability Test

In addition to the obtainment methods introduced above, reliability tests can also be used as input for failure rates in FMEA. These tests include reliability verification tests and accelerated life tests. You can find more information in reliability textbooks (Reliability Engineering (second edition) by E.A. Elsayed, etc.).

7.4.4 Mathematical Modeling

For the Electrical Wire Interconnection System (EWIS) component, such as wires and cables, the reliability data are not easy to obtain. In such cases, mathematical modeling methods can be useful.

The reliability data of wires and cables are influenced by various factors, such as the material, length, installation, and environment. By considering these factors, the mathematical modeling method can be used to obtain the reliability data of EWIS related components, such as wires and cables.

7.5 CASE STUDY OF ELECTRICAL POWER SYSTEM

This section analyses the failure modes and effects of Ram Air Turbine Generator Control Unit (RAT GCU) in emergency power supply system, in order to determine the possible failure modes of RAT GCU components and the effects on parts of the system, the whole system or even the aircraft. This FMEA is performed in supplier's view, while the assistances from aircraft manufacturer is also necessary in this process. For example, the failure effect to aircraft and failure identi-fication action are needed to be carefully discussed with aircraft manufacturer.

This case is simplified, and due to the differences in specific projects, the analysis and the results of this FMEA case are not necessarily applicable to other FMEAs. So, this case is for reference only.

7.5.1 Analysis Description

7.5.1.1 Failure Modes and Effects Analysis Description

In this case, the RAT GCU of the emergency power supply system is selected as the analysis object in accordance with the requirements of the FMEA procedure documents.

1. Requirement Analysis

   Analyze and identify the classification of the RAT GCU failures and their effects on aircraft and personnel on board and figure out the potential failures in accordance with the FMEA analysis requirements of the emergency power system. Determine if the recommended RAT GCU designs meet the requirements of the safety design of emergency power supply system according to the analysis result and summit base case data to the FTA of the system, to support the SSA of the AC power supply system.

2. Flight Phases

   The flight phases is the same as in Chapter 3.

3. FMEA Number

   According to ATA number rule, the number for power system is 24, of which the subnumber for emergency power supply system is 23. The FMEA of RAT GCU is numbered under the reference of this rule.

4. The Analyzed Components

   The list of components involved in the RAT GCU of emergency power supply system analyzed in this paper is as follows (Table 7.3).

5. The principle to identify system failure in the analysis

   A system, equipment, or component is considered to be failed if it is unable to perform the expected function. Failure mode and failure rate data are mainly from engineering experience (Table 7.4).

Table 7.3 Component list

Item #	Component number	Description
24-23-01-1	c1.1	RAT GCU
24-23-01-2	c1.2	Wires (used for providing RAT GEN with the excitation control signals that transmitted from RAT GCU)
24-23-01-3	c1.3	Wires (used as the power transfer between RAT GCU and RAT GEN heater)
24-23-01-4	c1.4	Wires (used as the connection between the Permanent Magnet Alternator of RAT GEN and RAT GCU)
24-23-01-5	c1.5	Wires (used for connecting the POR of RAT GEN and the POR of RAT GCU)
24-23-01-6	c1.6	Wires (used for connecting RAT_GCU_BIT of RAT GCU and the RAT HEATER BIT of DCU2)
24-23-01-7	c1.7	Wires (used for connecting RAT_GCU_BIT of RAT GCU and the RAT GCU BIT of DCU2)
24-23-01-8	c1.8	Wires (used for connecting the RAT RESEST SW and the RAT RESET of RAT GCU)

Table 7.4 Data of failure rate

Component	Component count	Failure rate of single component	Failure mode	Probability of failure mode	Data source
RAT GCU	1	3.521E−6/FH	The incorrect transmission of RAT HEATER FAIL message resulted by RAT GCU.	1.1E−7/FH	Reliability database
			RAT GCU reset function fail	1.2E−7/FH	Reliability database
			RAT GCU electrical protector to RAT generator fail (BIT is detectable)	1.9E−7/FH	Reliability database
			RAT GCU electrical protector to RAT generator fail (BIT is undetectable)	0.01E−7/FH	Reliability database
			RAT generator fails to achieve overload output due to RAT GCU failure.	0.95E−7/FH	Reliability database
			RAT generator failure due to RAT GCU failure (BIT is undetectable)	4.9E−7/FH	Reliability database
			RAT generator failure due to RAT GCU failure (BIT is detectable)	5.1E−7/FH	Reliability database
			BIT functional failure of RAT GCU	4.9E−7/FH	Reliability database
			The performance of RAT GCU is slightly affected but still within the scope the system stipulated	5.1E−7/FH	Reliability database
			FPGA failure caused by Single event upset failure	10E−7/FH	Reliability database
			Original random physical failure of FPGA	0.05E−7/FH	Reliability database

Note:
1. The failure rate of wire is calculated by mathematical modeling method.
2. All data has been adjusted from original database on purpose.

The hazard rating of the failure effects is determined by the power system function hazard analysis and the impact of the RAT GCU failure on the emergency AC power system and aircraft power supply.

6. Design Assumptions

The following design information used in FMEA of the emergency power supply system are assumed at the initial phase of the development; however, those information should be confirmed after the final design configuration has been determined:

1. The service life of aircraft: 50,000FH
2. The inspection interval of the release function and the power supply function of the aircraft RAT system: 2000FH
3. The inspection interval of the aircraft RAT system: 10 FH (as design requirement, the inspection is performed every day, and the mission time of aircraft is designed to be 10 FH each day).
4. The average flight duration of the aircraft is 1.2 h.

7.5.2 Ram Air Turbine Generator Control Unit Failure Modes and Effects Analysis Worksheet for AC Distribution System

To show the analysis process of RAT GCU FMEA, two representative failure modes are discussed below:

1. The incorrect transmission of RAT HEATER FAIL message resulted from RAT GCU (FMEA#24-23-01-1.1)

    According to history experience, "the incorrect transmission of RAT HEATER FAIL message resulted from RAT GCU failure" is a failure mode that is resulted from "RAT GCU circuit failure." By communicating with the supplier, it can be assured that the affected flight phase is on the ground. Based on RAT GCU function in the emergency power supply system the local effects of "the incorrect transmission of RAT HEATER FAIL message resulted from RAT GCU failure" will cause "RAT is incorrectly heated and cannot pass the BIT test." Since the incorrect transmission of RAT HEATER FAIL message has no safety impact on the upper level during the taxiing phase, the upper level and the final effect of this failure mode are "No effect." The failure modes indicate to the flight crew that the RAT HEATER FAIL message is displayed on the CMS page of the MFD, and other failure modes with the same indication include 24-23-01-3, 24-23-01-3.2 and RAT heating function failure, the failure isolation measures to be taken by the maintenance personnel are described in Fault Isolation Manual and the corrective action is to replace the RAT

GCU. The aircraft should not be dispatched with this failure mode. According to the internal inspection interval of the aircraft RAT system the exposure time is determined as 10FH, and the occurrence probability of the failure mode is 1.1E−6. The classification of this failure mode is identified as "No Safety Effect" according to EPS SFHA.

2. Wire short-circuit failure (FMEA#24-23-01-2.1)

As the subpart H was added to the FAR Part-25, which specify EWIS airworthiness requirements, the safety assessment process began to address the problem of wire failure. The wires of former aircraft type are generally considered not to fail, so the failure of wire was not normally analyzed in the FMEA. However, the failure modes and rates of the wire are now all analyzed in FMEA according to the requirements of the newly established 25.1709. The following is a description of the general process of the FMEA of the wire, taking an example of the wire that transmits RAT GCU excitation signal to the RAT generator. The main failure modes for the wire are short failure and open failure. The flight phase of short failure is "All." Based on the wire function, it is possible to determine that the local effect of the "short failure" is "Fails to transmit excitation signal." This effect is transmitted to the "It is unable for RAT generator to output power," and the final effect is "The aircraft power redundancy of aircraft is reduced." The corrective action for this failure mode is to replace the RAT GCU. When this failure mode occurs, the aircraft should not be dispatched with failure. The wire failure rate model calculates the failure rate of each wire. According to the release function of the RAT system and the power supply function check interval time, the exposure time is determined as 2000FH, and the occurrence probability of the failure mode is 3.8E−4. The classification of the failure mode is determine as "Minor" according to the EPS SFHA.

Finally the FMEA worksheet of the RAT GCU is summarized (Table 7.5).

As discussed in Section 7.3, FMES is a summary of the FMEA results which can provide failure rate inputs for the FTA basic events. Take FMEA# 24-23-01-1.5 and 24-23-01-1.9 as example, their effects on component are "the ability of controlling power supply quality of RAT generator decreases." The sum of failure rate is 6.05E−7/FH, then exposure time and classification are same as in FMEA. Finally the FMES worksheet of the RAT GCU is summarized in Table 7.6, and the corresponding FTAs are shown in Chapter 5.

Table 7.5 Worksheet of RAT GCU FMEA

	System: EPS			Component: RAT GCU				No. of the component: c1.1			
	Subsystem: emergency power supply system			Function of the component: control and protection of RAT generator				No. and version of the drawing:			
FMEA#	Failure mode/ cause	Flight phase	Failure effect a. Local effect b. Higher level effect c. Final effect (on aircraft)	Identification and corrective action of failures a. Indication to the flight crew b. Other failures with the same indication c. Identification, isolation, and corrective actions of failures taken by the flight crew d. The effects of possible inappropriate actions e. Failure isolation— maintenance personnel f. Corrective action— maintenance personnel	Dispatch requirements with failures a. If able, the aircraft can be dispatched b. If able, what is the flight limitation	Component failure rate (1E−7/ FH)	Mode failure rate (1E−7/ FH)	Exposure time (FH)	Mode probability	Classification	Remarks
24-23-01-1.1	The incorrect transmission of RAT HEATER FAIL message resulted from RAT GCU Failure cause: RAT GCU circular failure	G	a. RAT is incorrectly heated and cannot pass the BIT test b. No effect c. No effect	a. MFD CMS page displays: RAT HEATER FAIL b. 24-23-01-3.1, 24-23-01-3.2 and RAT heating function failure c. None d. None e. Replace RAT GCU f. Replace RAT GCU	a. Disable b. N/A	35.21	1.1	10	1.1E−6	No Safety Effect	Hidden failure
24-23-01-1.2	The reset function failure of RAT GCU. Cause: RAT GCU circular failure	T, F1~ F4, L	a. The RAT generator contactor reset cannot be implemented as required (e.g., when the RAT generator is overloaded). b. RAT generator fails to supply power. c. The safety redundancy of the aircraft power supply system is reduced.	a. None b. None c. None d. None e. Replace RAT GCU f. Replace RAT GCU	a. Disable b. N/A	35.21	1.2	2000	2.4E−4	No Safety Effect	Hidden failure

ID	Function failure	Phase	Failure effect	Detection / Action	Compensating	Rate	Ratio	MTBF	Probability	Severity	Remarks
24-23-01-1.3	The function failure of electrical protection to RAT generator by RAT GCU (BIT is detectable). Failure cause: RAT GCU circuit failure.	T, F1 ~ F4,L	a. It is unable for RAT GCU to protect the RAT generator from electrical failure. b. In the case of emergency power supply, if the RAT generator electrical failure occurs, the electrical protection cannot be performed. c. The quality of emergency power supply is reduced, or overheat and fire is caused by high current failure.	a. MFD CMS page displays: RAT GCU/ WRG FAIL. b. 24-23-01-1.7; 24-23-01-1.8; 24-23-01-7.1; 24-23-01-7.2 c. None d. None e. See 24-23-00-110-101 in Fault Isolation Manual f. Replace RAT GCU	a. Disable b. N/A	35.21	1.9	10	1.9E−6	No Safety Effect	Hidden failure
24-23-01-1.4	The function failure of electrical protection to RAT generator by RAT GCU (BIT is undetectable) Failure cause: RAT GCU circuit failure.	T, F1 ~ F4, L	a. It is unable for RAT GCU to protect the RAT generator from electrical failure. b. In the case of emergency power supply, if the RAT generator electrical failure occurs, the electrical protection cannot be performed, and there is an damage on RAT generator. c. No effect.	a. None b. None c. None d. None e. Replace RAT GCU f. Replace RAT GCU	a. Able b. No limit	35.21	0.01	2000	2E−6	No Safety Effect	Hidden failure
24-23-01-1.5	RAT GCU loses the ability to control RAT generator to conduct the overload output. Failure Cause: RAT GCU circuit failure.	T, F1 ~ F4, L	a. RAT generator cannot achieve the overload output b. In the need of overload output, the capacity of emergency power supply is insufficient and, power supply quality declines. c. The safety redundancy of aircraft power supply system is reduced.	a. None b. None c. None d. None e. Replace RAT GCU f. Replace RAT GCU	a. Disable b. N/A	35.21	0.95	2000	1.9E−4	Minor	Hidden failure

(Continued)

Table 7.5 (Continued)

	System: EPS			Component: RAT GCU					No. of the component: c1.1
	Subsystem: emergency power supply system			Function of the component: control and protection of RAT generator					No. and version of the drawing:
24-23-01-1.6	RAT GCU loses the control to RAT generator due to function failure. (BIT is undetectable) Failure Cause: RAT GCU circuit failure.	T, F1~F4, L	a. It is unable to control the RAT generator. b. RAT generator fails to supply power. c. The power supply redundancy of aircraft is reduced.	a. None b. None c. None d. None e. Replace RAT GCU f. Replace RAT GCU	35.21	4.9	2000	9.8E–4	Minor
24-23-01-1.7	RAT GCU loses the control to RAT generator due to function failure. (BIT is detectable) Failure cause: RAT GCU Circuit failure	T, F1~F4, L	a. The RAT generator cannot be controlled and cannot pass the BIT test. b. The power system loss a AC power supply. c. The power redundancy of aircraft is reduced.	a. MFD CMS page displays: RAT GCU/ WRG FAIL b. 24-23-01-1.3; 24-23-01-1.8; 24-23-01-7.1; 24-23-01-7.2 c. None d. None e. See 24-23-00-110-101 in the Fault Isolation Manual f. Replace RAT GCU	35.21	5.1	10	5.1E–6	Minor
24-23-01-1.8	BIT function failure of RAT GCU Failure cause: RAT GCU Circuit failure	G	a. RAT GCU fail to pass BIT b. No effect. c. No effect.	a. MFD CMS page displays: RAT GCU/ WRG FAIL b. 24-23-01-1.3; 24-23-01-1.7; 24-23-01-7.1; 24-23-01-7.2 c. None d. None e. See 24-23-00-110-101 in the Fault Isolation Manual f. Replace RAT GCU.	35.21	4.9	10	4.9E–6	No Safety Effect

System: power system

Subsystem: emergency power supply system   Component: wire

Function of the component: transmits RAT GCU excitation signal to the RAT generator

No. of the component: c1.2

No. and version of the drawing:

FMEA#	Failure mode/ cause	Flight phase	Failure effect a. Local effect b. Higher level effect c. Final effect (on aircraft)	Identification and corrective action of failures a. Indication to the flight crew b. Other failures with the same indication c. Identification, isolation, and corrective actions of failures taken by the flight crew. d. The effects of the possible inappropriate actions	Dispatch requirements with failures a. If able, the aircraft can be dispatched b. If able, what is the flight limitation	The failure rate of component (1E−7/FH)	Mode failure rate 1E−7/FH	Exposure time (FH)	Mode probability	Classification	Hidden failure	Remarks
24-23-01-1.9	RAT GCU performance slightly decrease. Failure Cause: RAT GCU circuit failure	T, F1~F4, L	a. The control performance of RAT GCU on RAT generator decrease slightly. b. The quality of RAT electrical output might decrease slightly. c. No effect.	a. None b. None c. None d. None e. Replace RAT GCU f. Replace RAT GCU	a. Able b. No limit	35.21	5.1	2000	1.02E−3	No Safety Effect		
24-23-1-1.10	FPGA failure caused by single-event effect	All	a. The RAT GCU excitation control fails. b. N/A c. N/A	a. N/A b. N/A c. N/A d. N/A e. Reset RAT GCU in flight f. Reset RAT GCU in flight	a. Able b. After reset, it should pass BIT		12	1.2	1.44E−6	Hazardous	Reference to Chapter 9	
24-23-1-1.11	Original random physical failure of FPGA	All	a. The RAT GCU excitation control fails. b. N/A c. N/A	a. N/A b. N/A c. N/A d. N/A e. Replace RAT GCU f. Replace RAT GCU	a. Disable b. N/A	0.05	0.05	1.2	6E−9	Hazardous	Reference to Chapter 9	

(Continued)

Table 7.5 (Continued)

System: power system — **Component: wire** — **No. of the component: c1.2**

Subsystem: emergency power supply system

Function of the component transmits RAT GCU excitation signal to the RAT generator

FMEA#	Flight Phase	Failure effect	Dispatch requirements with failures	Identification and corrective action of failures	The Failure Rate of Component (1E-7/FH)	Mode Failure Rate 1E-7/FH	Exposure time (FH)	Mode Probability	Classification	Remarks
24-23-01-2.1	All	Short failure. Failure cause: Short occurs between wires. a. Fails to transfer transmit signal. b. It is unable for RAT generator to output power. c. The aircraft power redundancy of aircraft is reduced.	a. Disable b. N/A	a. None b. None c. None d. None e. Replace wires (Failures can be found through detecting the power supply function of RAT generator.) f. Replace wires; e. Failure isolation—maintenance personnel f. Corrective action—maintenance personnel	3.1	1.9	2000	3.8E−4	Minor	Hidden failure
24-23-01-2.2	All	Wires open failure. Failure cause: open. a. Fails to transfer transmit signal. b. It is unable for RAT generator to output nor transfer the power. The power system loses one AC power. c. The power supply redundancy of aircraft is reduced.	a. Disable b. N/A	a. None b. None c. None d. None e. Replace wires Failures can be found through detecting the power supply function of RAT generator. f. Replace wires	3.1	1.2	2000	2.4E−4	Minor	Hidden failure

System: power system — **Component: wires** — **No. of the component: c1.3**

Subsystem: emergency power supply system

Function of the component: RAT GCU heating power output

FMEA#	Flight Phase	Failure effect	Dispatch requirements with failures	Identification and corrective action of failures	The Failure Rate of Component (1E-7/FH)	Mode Failure Rate 1E-7/FH	Exposure time (FH)	Mode Probability	Classification	Remarks
Failure mode/ cause	Flight Phase	a. Local effect b. Higher level effect c. Final effect (on aircraft)	a. If able, the aircraft can be dispatched b. If able, what is the flight limitation	a. Indication to the flight crew b. Other failures with the same indication c. Identification, isolation, and corrective actions of failures taken by the flight crew d. The effects of the possible inappropriate actions						

FMEA#	Failure mode/cause	Flight phase	Failure effect: a. Local effect b. Higher level effect c. Final effect (on aircraft)	Identification and corrective action of failures: a. Indication to the flight crew b. Other failures with the same indication c. Identification, isolation, and corrective actions of failures taken by the flight crew d. The effects of the possible inappropriate actions e. Failure isolation—maintenance personnel f. Corrective action—maintenance personnel	Dispatch Requirements with Failures: a. If able, the aircraft can be dispatched b. If able, what is the flight limitation	The Failure Rate of Component (1E−7/FH)	Mode Failure Rate 1E−7/FH	Exposure time (FH)	Mode Probability	Classification	Remarks
24-23-01-3.1	Short failure of wires. Failure cause: wire short.	All	a. It is unable for RAT GCU to provide heat for RAT generator. b. The RAT generator may fail due to losing the heating input. c. The power supply redundancy of aircraft is reduced.	a. MFD CMS page displays: RAT HEATER FAIL. b. RAT heating function failure: 24-23-01-1.1; 24-23-01-3.2; 24-23-01-6.1; 24-23-01-6.2 c. None d. None e. See 24-23-00-110-102 in the Fault Isolation Manual f. Replace RAT GCU	a. Disable b. N/A	3.1	1.9	10	1.9E−6	Minor	Hidden failure
24-23-01-3.2	Short failure of wires. Failure cause: open of wires.	All	a. It is unable for RAT GCU to provide heat for RAT generator. b. The RAT generator may fail due to losing the heating input. c. The power supply redundancy of aircraft is reduced.	a. MFD CMS page displays: RAT HEATER FAIL. b. RAT heating function failure: 24-23-01-1.1; 24-23-01-3.1; 24-23-01-6.1; 24-23-01-6.2 c. None d. None e. See 24-23-00-110-102 in the Fault Isolation Manual f. Replace RAT GCU	a. Disable b. N/A	3.1	1.2	10	1.2E−6	Minor	Hidden failure

System: power system

Subsystem: emergency power supply system

Component wires

Function of the component: the PMG point and RAT generator's PMG point that connects RAT GCU

No. of the component: c1.4

No. and version of the drawing:

(Continued)

Table 7.5 (Continued)

System: power system | **Component: wires** | **No. of the component: c1.4**

Subsystem: emergency power supply system | Function of the component: the PMG point and RAT generator's PMG point that connects RAT GCU | No. and version of the drawing:

FMEA#	Failure mode/cause	Flight phase	Failure effect a. Local effect b. Higher level effect c. Final effect (on aircraft)	Identification and corrective action of failures	The Failure Rate of Component (1E−7/FH)	Mode Failure Rate 1E−7/FH	Exposure time (FH)	Mode Probability	Classification	Remarks
24-23-01-4.1	Short failure. Failure cause: Short occurs between wires.	All	a. It is unable for RAT GCU to control RAT generator b. It is unable for RAT generator to output power c. The power supply redundancy of aircraft is reduced.	a. None b. None c. None d. None e. Replace wires (Failures can be found through detecting the power supply function of RAT generator.) f. Replace wires e. Failure isolation—maintenance personnel f. Corrective action—maintenance personnel	6.2	3.8	2000	7.6E−4	Minor	Hidden failure
24-23-01-4.2	Wires open failure. Failure cause: open	All	a. It is unable for RAT GCU to control RAT generator b. It is unable for RAT generator to output power. c. The power supply redundancy of aircraft is reduced.	a. None b. None c. None d. None e. Replace wires (Failures can be found through detecting the power supply function of RAT generator.) f. Replace wires	6.2	2.4	2000	4.8E−4	Minor	Hidden failure

System: power system | **Component: RAT GCU** | **No. of the component: c1.5**

Subsystem: emergency power supply system | Function of the component: transmitting regulation voltage to RAT GCU | No. and version of the drawing:

FMEA#	Failure mode/ cause	Flight phase	Failure effect a. Local effect b. Higher level effect c. Final effect (on aircraft)	Identification and corrective action of failures a. Indication to the flight crew b. Other failures with the same indication	Dispatch Requirements with Failures a. If able, the aircraft can be dispatched	The Failure Rate of Component (1E−7/FH)	Mode Failure Rate 1E−7/FH	Exposure time (FH)	Mode Probability	Classification	Remarks

FMEA#	Failure mode/cause	Flight phase	Failure effect: a. Local effect b. Higher level effect c. Final effect (on aircraft)	Identification and corrective action of failures: a. Indication to the flight crew b. Other failures with the same indication c. Identification, isolation, and corrective actions of failures taken by the flight crew d. The effects of the possible inappropriate actions e. Failure isolation—maintenance personnel f. Corrective action—maintenance personnel	Dispatch requirements with failures: a. If able, the aircraft can be dispatched b. If able, what is the flight limitation	The failure rate of component (1E−7/FH)	Mode failure rate (1E−7/FH)	Exposure time (FH)	Mode probability	Classification	Remarks
24-23-01-5.1	Short failure. Failure cause: Short occurs between wires.	All	a. RAT GCU electric protective tripping and RAT generator excitation. b. RAT generator is not able to output the electric power. It lost one AC power. c. Flight power supply decrease.	a. None b. None c. None d. None e. Replace wires (RAT generator power supply function detection will find the failure.) f. Replace wires	a. Disable b. N/A	6.2	3.8	2000	7.6E−4	Minor	Hidden failure
24-23-01-5.2	Wires open failure. Failure cause: Wires open.	All	a. RAT GCU electric protective tripping and RAT generator excitation. b. Generator is not able to output the electric power. It lost one AC power. c. Flight power supply decrease.	a. None b. None c. None d. None e. Replace wires (RAT generator power supply function detection will find the failure.) f. Replace wires	a. Disable b. N/A	6.2	2.4	2000	4.8E−4	Minor	Hidden failure

System: power system

Subsystem: emergency power supply system

Component: wires

Function of the component: transfer RAT heater BIT signal

No. of the component c1.6

No. and version of the drawing:

(Continued)

Table 7.5 (Continued)

System: power system		Component: wires		No. of the component c1.6
Subsystem: emergency power supply system		Function of the component: transfer RAT heater BIT signal		No. and version of the drawing:

				d. The effects of the possible inappropriate actions e. Failure isolation—maintenance personnel f. Corrective action—maintenance personnel	b. If able, what is the flight limitation						
24-23-01-6.1	Short failure. Failure cause: Short	G	a. Wires failure effect RAT GCU BIT. b. RAT GCU is not able to pass BIT. c. No effect.	a. MFD CMS page displays: RAT HEATER FAIL. b. RAT heater function failure: 24-23-01-1.1; 24-23-01-3.1; 24-23-01-3.2; 24-23-01-6.2 c. None d. None e. See 24-23-00-110-102 in the Fault Isolation Manual f. Replace RAT GCU	a. Disable b. N/A	59	38	10	3.8E−5	No Safety Effect	Hidden failure
24-23-01-6.2	Wires open failure. Failure cause: open	G	a. Wires failure effect RAT GCU BIT. b. RAT GCU is not able to pass BIT. c. No effect.	a. MFD CMS page displays: RAT HEATER FAIL. b. RAT heater function failure: 24-23-01-1.1; 24-23-01-3.1; 24-23-01-3.2; 24-23-01-6.1 c. None d. None e. See 24-23-00-110-102 in the Fault Isolation Manual f. Replace RAT GCU	a. Disable b. N/A	59	21	10	2.1E−5	No Safety Effect	Hidden failure

System: power system

Component: wires

No. of the component: c1.7

Subsystem: emergency power supply system

Function of the component RAT GCU BIT signal transferring

No. and version of the drawing:

FMEA#	Failure mode/cause	Flight phase	Failure effect a. Local effect b. Higher level effect c. Final effect (on aircraft)	Identification and corrective action of failures a. Indication to the flight crew b. Other failures with the same indication c. Identification, isolation, and corrective actions of failures taken by the flight crew d. The effects of the possible inappropriate actions e. Failure isolation—maintenance personnel f. Corrective action—maintenance personnel	Dispatch Requirements with Failures a. If able, the aircraft can be dispatched b. If able, what is the flight limitation	The Failure Rate of Component (1E-7/FH)	Mode Failure Rate (1E-7/FH)	Exposure time (FH)	Mode Probability	Classification	备注 Remarks
24-23-01-7.1	Short failure. Failure cause: Short occurs between wires.	G	a. Wires failure effect RAT GCU BIT. b. RAT GCU is not able to pass BIT. c. No effect.	a. MFD CMS page displays: RAT GCU/WRG FAIL. b. 24-23-01-1.3; 24-23-01-1.7; 24-23-01-1.8; 24-23-01-7.2 c. None d. None e. Replace RAT GCU	a. Disable b. N/A	59	38	10	3.8E-5	No Safety Effect	Hidden failure
24-23-01-7.2	Wires open failure. Failure cause: open	G	a. Wires failure effect RAT GCU BIT. b. RAT GCU is not able to pass BIT. c. No effect.	a. MFD CMS page displays: RAT GCU/WRG FAIL b. 24-23-01-1.3; 24-23-01-1.7; 24-23-01-1.8; 24-23-01-7.1 c. None d. None e. See 24-23-00-110-101 in the Fault Isolation Manual f. Replace RAT GCU	a. Disable b. N/A	59	21	10	2.1E-5	No Safety Effect	Hidden failure

System: power system

Component: wires

No. of the component: c1.8

Subsystem: emergency power supply system

Function of the component: Transferring the reset control signal from RAT reset switch

No. and version of the drawing:

FMEA#	Failure mode/cause	Flight phase	Failure effect	Identification and corrective action of failures	Dispatch requirements with failures	The failure rate of component (1E−7/FH)	Mode failure Rate (1E−7/FH)	Exposure time (FH)	Mode probability	Classification	Remarks
			a. Local effect b. Higher level effect c. Final effect (on aircraft)	a. Indication to the flight crew b. Other failures with the same indication c. Identification, isolation, and corrective actions of failures taken by the flight crew. d. The effects of the possible inappropriate actions e. Failure isolation—maintenance personnel f. Corrective action—maintenance personnel	a. If able, the aircraft can be dispatched b. If able, what is the flight limitation						
24-23-01-8.1	Short failure.Failure cause: Short occurs between wires.	All	a. RAT generator reset function failure b. No effect c. No effect	a. None b. None c. None d. None e. Replace wires f. Replace wires	a. Disable b. N/A	41	22	2000	4.4E−3	No Safety Effect	Hidden failure
24-23-01-8.2	Wires open failure. Failure cause: open	All	a. RAT generator reset function failure b. No effect c. No effect	a. None b. None c. None d. None e. Replace wires f. Replace wires	a. Disable b. N/A	41	19	2000	3.8E−3	No Safety Effect	Hidden failure

Note:
1. This failure is only caused by SEE and has no relationship with FPGA quality, so there is only mode failure rate.

Table 7.6 Worksheet of RAT GCU FMES

System: EPS   No. of ATA: 24-23

FMES#	Failure mode	Failure rate (1E−7/FH)	Exposure time (FH)	Classification	FMEA#
24-23-01	BIT function failure of RAT GCU	4.9	10	No Safety Effect	24-23-01-1.8
24-23-02	RAT GCU failure and could not pass the built-in testing	126.1	10	No Safety Effect	24-23-01- 1.1, 24-23-01-1.3, 24-23-01-1.7, 24-23-01-6.1, 24-23-01-6.2, 24-23-01-7.1, 24-23-01-7.2
24-23-03	RAT generator fails to supply power caused by RAT GCU failure	20.4	2000	Minor	24-23-01-1.6, 24-23-01-2.1, 24-23-01-2.2, 24-23-01-4.1, 24-23-01-4.2, 24-23-01-5.1, 24-23-01-5.2
24-23-04	The function failure of electrical protection and reset to RAT generator by RAT GCU	42.21	2000	No Safety Effect	24-23-01-1.2, 24-23-01-1.4, 24-23-01-8.1, 24-23-01-8.2
24-23-05	The ability of controlling power supply quality of RAT generator decreases	6.05	2000	No Safety Effect	24-23-01-1.5, 24-23-01-1.9
24-23-06	RAT loss heat protection	3.1	10	Minor	24-23-01-3.1, 24-23-01-3.2
24-23-07	The RAT GCU excitation control fails	12.05	1.2	Hazardous	24-23-01-1.10, 24-23-01-1.11

REFERENCES

[1] SAE ARP 4761. Guidelines and methods for conducting the safety assessment process on civil airborne systems and equipment. SAE International; 1996.

[2] MIL-HDBK-217F Notice-2. Reliability prediction handbook for electronic equipment. U.S. Department of Defense; 1995.

[3] PRISM. Reliability prediction and database for electronic and non-electronic parts. Reliability Analysis Center; 2000.

[4] FIDES-2009A. Methodologie de fiabilite les systemes elctroniques. FIDES Group; 2009.

CHAPTER 8

System Safety Assessment

Contents

8.1 INTRODUCTION

System Safety Assessment (SSA) is a systematic and comprehensive assessment on the architecture, design, and installation of the systems to ensure that relevant safety requirements are met.

During the SSA process, all critical failure conditions and their effects on aircraft will be assessed as similar approach to PSSA. However, Preliminary System Safety Assessment (PSSA) and SSA vary in scope and purpose. PSSA is aimed at allocating system safety requirements to items in a top-down way, described in Chapter 5, Preliminary System Safety Assessment. SSA focuses on verifying whether the design can meet the qualitative and quantitative safety requirements from Functional Hazard Assessment (FHA) and PSSA in a bottom–up way. SSA is a continuous and iterative process throughout the aircraft development process. For each PSSA implemented at different levels, there is a corresponding SSA. System level SSA is to verify the safety requirements in Aircraft Functional Hazard Assessment (AFHA) and/or System Functional Hazard Assessment (SFHA) or PSSA.

The objectives of SSA include [1]:

- Verifying whether the safety requirements (design requirements) and objectives in the SFHA are met.
- Verifying whether the safety-related requirements derived from the design of the system architecture, equipment, software, and aircraft installation are met.
- Confirming whether all necessary supporting materials identified in FHA/PSSA have been completed.

8.2 SYSTEM SAFETY ASSESSMENT INPUTS

As shown in Fig. 8.1, the inputs required by the SSA process can be divided into two categories:

1. Top-level system safety requirements and design

   The "top-level system safety requirements" refers to the failure conditions and their safety effects identified in the SFHA, as well as the safety requirements from the PASA or higher level PSSA. During the SSA process, these requirements will be verified as primary safety objectives. The "safety design" includes system function, architecture description, design principles, and the interactions with the adjacent systems. Safety design sets up the foundation for SSA assessment.

2. The safety-related requirements and supporting materials derived from the system design and the FHA/PSSA process

   The "safety-related requirements" include derived requirements, Development Assurance Level (DAL) requirements, independence requirements, as well as operational and maintenance requirements, and so on, from the PSSA process. The "supporting materials" mainly contains the Common Cause Analysis (CCA) data (including Zonal Safety Analysis, Particular Risk Analysis, and Common Mode Analysis), Failure Modes and Effects Analysis (FMEA)/Failure Modes and Effects Summary (FMES) from equipment suppliers, validation and verification results, research and analysis, operational and maintenance limitations. All of these will provide support for demonstrating compliance with safety objectives in the SSA process.

Figure 8.1 SSA process. *SSA*, System Safety Assessment.

8.3 SYSTEM SAFETY ASSESSMENT PROCESS

Based on the above inputs, the activities in the SSA mainly contains

1. Verification of the top-level safety requirements of failure conditions in the SFHA, including the classification validation of failure conditions, followed by the verification of top-level safety requirements.

2. Verification of the safety-related requirements from PSSA, including verification of derived requirements, confirmation of the compliance with the DAL requirements, confirmation of the maintenance means for the hidden failures and Candidate Certification Maintenance Requirement (CCMR) items, confirmation of the operational procedures and limitations derived from the safety assessment, as well as the result checking of CCA and Master Minimum Equipment List (MMEL).

3. At the end of the development process of the system, all of the safety data should be checked to ensure traceability between safety assessment activities.

Note: According to the SAE ARP4754A [2], the "safety requirements" refers to minimum performance constraints for both availability and integrity of the function. The "safety-related requirements" are typically the independence requirements, probabilistic availability and integrity requirements, no single failure criteria, monitor performance requirements, safety or protective features, DAL, and operational and maintenance limitations.

8.3.1 Verification of Top-Level Safety Requirements of Failure Conditions in System Functional Hazard Assessment

8.3.1.1 Confirmation of Validating the Classification of Failure Conditions

Prior to the verification of the safety requirements of the failure conditions, the classification should be assessed to ensure the validity of subsequent verification results. For failure conditions whose failure effects are difficult to determine, classification of the failure conditions should be validated by providing supporting materials.

As for hazardous failure conditions, the engineering simulator test, flight simulator test, "iron bird" test or ground test can be used to validate the classification. If the test cannot be carried out, an analysis report should be provided.

As for the major and minor failure conditions, the engineering simulator test, flight simulator test, "iron bird" test, ground test, or flight test can be used to validate the classification. If a pretest analysis is necessary, an analysis report should be provided.

8.3.1.2 Verification of the Safety Requirements of Failure Conditions

A qualitative and quantitative Fault Tree Analysis (FTA) was performed by taking failure conditions as the top events in combination with the system architecture. The analysis method is similar to PSSA, except that during the PSSA process, the engineers implement the quantitative calculation with the reliability data of basic events from engineering experience to ensure that safety requirements allocated to the basic events are practicable; however, during the SSA process, the fault tree calculation is done with reliability data of components from FMEA/FMES to verify if the product's design satisfies the safety requirements.

8.3.2 Verification of the Safety-Related Requirements From Preliminary System Safety Assessment

8.3.2.1 Verification of the Derived Safety-Related Requirements

According to Chapter 5, Preliminary System Safety Assessment, there will be a large number of derived safety-related requirements in the PSSA process, such as independence requirements, quantitative and qualitative requirements from adjacent systems/equipment, and relevant design requirements. These derived requirements will be verified through analysis, calculation, test, or other methods. In particular, some of the independence requirements should be verified by CCA, described in Chapter 6, Common Cause Analysis.

> Note: According to the SAE ARP4754A [2], derived requirements refer to additional requirements resulting from design or implementation decisions during the development process that are not directly traceable to higher level requirements.

8.3.2.2 Confirmation of the Maintenance Means for Hidden Failures and Candidate Certification Maintenance Requirement Items

According to Chapter 5, Preliminary System Safety Assessment, hidden failures are identified and the maintenance intervals are determined during the PSSA process. Therefore, during the SSA process, it is necessary to further evaluate whether the maintenance interval can meet the safety requirements based on the detailed system architecture and updated fault trees. In addition, it is necessary to check whether the hidden failures and maintenance tasks contained in the fault trees of catastrophic and hazardous failure conditions have been submitted to the CMCC as a CCMR item or other maintenance items. The CMCC coordinates the CCMR items and the MSG-3 maintenance activities to finalize the CMR items and maintenance interval recorded in Appendix 1 or Appendix A of the Maintenance Review Board Report (MRBR).

8.3.2.3 Confirmation of the Compliance with Development Assurance Level Requirements

After assigning Function Development Assurance Levels (FDALs) to the functions of the systems and IDALs to the items in PSSA, the SSA process is required to confirm the compliance with the development assurance requirements for system functions/airborne software/complex airborne electronic hardware according to SAE ARP4754A [2], RTCA DO-178C [3], and RTCA DO-254 [4], respectively. Traceability between requirements and compliance substantiations should be established.

8.3.2.4 Confirmation of the Operational Procedures and Limitation

Operational procedures and limitations may have significant effects on the classification of the failure conditions and, finally, on operation safety. Therefore, during the SSA process, it is necessary to check that operational limitations and procedures derived from safety analysis have been verified and recorded in the relevant documents (such as the AFM), including rules for display, alarm, flight crew operation, etc.

8.3.2.5 Correctness Checks for Master Minimum Equipment List

Some failures of the airborne equipment or parts will not affect the safe operation of the aircraft. For example, the APU generator is the backup portion of the Electrical Power System (EPS). Due to the limitation of work conditions, it's usually not considered in the quantitative FTA. In other words, the loss of the APU generator has no effect on the results of the FTA. The aircraft with the failure still meets airworthiness requirements. We may call this kind of equipment or parts "GO items."

The operational staff requires efficient and reliable reference documents to quickly determine whether the failures affect flight safety and whether the aircraft can go with failures. This document is called MMEL. MMEL can ensure operation safety and reduce the cost of operations and equipment reservation.

For the safety assessment of MMEL items, the constructed fault trees may be different from those for a normal flight. We consider a particular flight for which you know that one or several GO items are failed at takeoff: to compute the probability of failure conditions containing the failed GO items, the analyst must "delete" all events related to GO items in the fault trees. Correctness checking is necessary to determine whether airworthiness requirements are still met for all the particular flights under the MMEL.

Note: Airworthiness authorities may additionally ask for compliance with other safety objectives specific to MMEL items. For instance, they may require safety objectives to be met for the "mean probability" of a failure condition with GO items.

8.3.3 Checking for the Traceability of Safety Data

At the end of the system development process, each SSA must review the traceability of all data related to the system safety to ensure the activities are coherent.

8.3.3.1 Establishment of the Safety Data Architecture

The safety-related data includes system requirements, functions, and design description; safety analysis data and supporting data, and operational and maintenance data derived from safety analysis.

- Requirements, functions, and design description data, providing necessary inputs for the safety analysis, include:
  - system/equipment requirements, functions data
  - system/equipment design description data (SDD)
  - system interface description data (SID)
- Safety analysis data and supporting data include:
  - system Certification Plan (CP)
  - AFHA data
  - PASA data
  - SFHA data
  - PSSA data
  - SSA data
  - FTA data
  - FMEA/FMES data
  - CCA data
  - supporting materials (e.g., test report for validation and verification, and DAL compliance materials)
- Operational and maintenance data includes:
  - MMEL data
  - CCMR data
  - AFM

8.3.3.2 Traceability Between System Safety Data

The relationship model between the safety data can be shown in Fig. 8.2.

L1: Traceability between CP and System Safety Data

The system safety assessment process aims to demonstrate that safety-related regulations in CP have been met. The SSA data are supposed to list the regulations and introduce the detailed means of compliance combining with the SSA process.

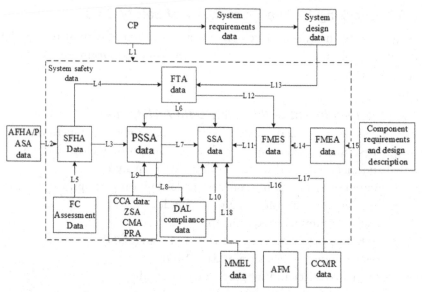

Figure 8.2 Traceability model of system safety data.

L2: Traceability between AFHA Data and SFHA Data

The traceability between AFHA and SFHA focuses on two points:
- consistency between aircraft-level functions and system-level functions;
- consistency between the safety effects and classification of aircraft-level failure conditions and that of system-level failure conditions.

L3: Traceability between SFHA Data and PSSA/SSA Data

The traceability between PSSA/SSA data and SFHA data focuses on two points:
- The PSSA/SSA analysis must be based on the results of the SFHA. The system safety objectives and requirements are determined during the SFHA process. In the PSSA/SSA analysis, the qualitative and quantitative FTA takes these safety requirements as the starting point (top event).
- The analysis depth of PSSA/SSA is based on SFHA results. For example, qualitative and quantitative safety assessments are required for the catastrophic, hazardous, and major failure conditions of highly integrated or complex systems. However, only a qualitative safety assessment is required for minor and no-safety-effect failure conditions.

L4: Traceability between SFHA Data and FTA Data

The system safety requirements are determined during the SFHA process. The qualitative and quantitative FTA in PSSA/SSA should take these safety requirements as the starting point (top event).

L5: Traceability between Failure Condition Classification Assessment Data and SFHA Data

According to Section 8.3.1.1, the corresponding validation activities are required for the classification of failure conditions in the SFHA. The validation data should be indexed in the SFHA data.

L6: Traceability between PSSA/SSA Data and FTA Data

Both safety requirements allocation and quantitative calculation in PSSA/SSA should be based on FTA. Therefore, it is necessary to index the corresponding FTA data.

L7: Traceability between SSA Data and PSSA Data

When safety requirement compliance is demonstrated, the PSSA data, especially the results of the DAL assigned, should be indexed in the SSA data.

L8: Traceability between DAL Compliance Data and PSSA Data

The DAL compliance data of the corresponding items should be identified and indexed in the PSSA.

L9: Traceability between PSSA/SSA Data and CCA Data

CCA data analyzes the AND-Gates events in the fault tree of the catastrophic or hazardous failure conditions to determine whether there is a single failure which can lead to the failure condition by itself. It is necessary to index the CCA data in the PSSA/SSA data.

L10: Traceability between SSA Data and DAL Compliance Data

DAL compliance data should be indexed in the SSA data based on the results of DAL allocation of items in the PSSA, which mainly includes the Plan of Software/Hardware Airworthiness Certification (PHAC/PSAC), the Software/Hardware Accomplishment Summary (SAS/HAS), the Software/Hardware Configuration Index (HCI/SCI), etc.

L11: Traceability between SSA Data and FMES Data

During the SSA process, the reliability data of the basic events in the fault trees, such as failure rates and exposure time, should be identified according to the corresponding FMES data. Therefore, FMES data should be traced and indexed in the SSA data.

L12: Traceability between FMES Data and FTA Data

To enable FMES to effectively support the quantitative FTA in SSA, FMES is required to consider the basic events of the fault trees as the objectives analyzed, so that it is easy to obtain the failure rates and exposure time.

L13: Traceability between FTA Data and System Design Data

The construction of the system fault trees is based on the system design and architecture. Therefore, the system design data, such as final version SID and SDD, should be indexed in the FTA data.

L14: Traceability between FMES Data and FMEA Data

FMES sorts and summarizes the relevant FMEA items in the form of considering SSA and FTA basic events as the targeted failure effects. It is necessary to establish the relationship between FMEA and FMES items in FMES data.

L15: Traceability between FMEA and Equipment Requirements Design Data

FMEA should be based on equipment requirements and design data to determine failure modes and safety-related failure effects. The equipment requirements and design description data should be indexed and traced in FMEA data.

L16: Traceability between SSA Data and AFM

The SSA summarizes the safety-related flight operational requirements in abnormal and emergency situations and verifies whether they are included in the AFM.

L17: Traceability between SSA Data and CCMR Data

The hidden failures and maintenance tasks related to catastrophic and hazardous failures conditions should be summarized in the SSA data, and verified whether they are contained in the CCMR data and Appendix I or A of MRBR.

L18: Traceability between SSA Data and MMEL

The proposed MMEL should be verified to determine whether it meets the minimum safety requirements from the regulations in the SSA data.

8.4 SYSTEM SAFETY ASSESSMENT OUTPUTS

The results of the SSA process should be documented and archived as follows such that there is traceability of the steps taken in developing the SSA data:

1. Failure condition quantitative analysis result summary, listing of failure conditions from the SFHA with quantitative analysis (i.e., FTA) results comparison.
2. Supporting documents used to assess the classification of each failure condition or safety requirement.
3. Safety requirement quantitative and qualitative analysis result summary, listing of safety requirements allocated or derived in the safety

assessment activities with verification results comparison, including operational procedures and limitation, CCMR items, MMEL correctness checking, etc.

4. Summary of DAL compliance, system, and items developed in accordance with assigned DAL.
5. CCA results summary (no independence issues).
6. Traceability checking result summary.

NOTE: However, the completion of the SSA doesn't mean the end of all the safety assessment activities. Besides, the system safety should be verified in the aircraft level by Aircraft Safety Assessment (ASA). Similar to SSA, ASA is a systematic, comprehensive evaluation of the aircraft to show that top-level safety requirements established by the PASA are met, which is described in Chapter 4, Aircraft Functional Hazard Assessment. According to the SAE ARP4754A [2], it should provide a summary of the aircraft safety activities from the beginning of the concept development to the completion of the detailed design development. It aims to show compliance with aircraft level requirements and objectives and give assurance that the appropriate methods and process have been applied. ASA is organized by the OEM rather than system suppliers. Because this book focuses on the safety assessment of the EPS system, aircraft-level safety synthesis isn't organized and shown in this book.

8.5 CASE STUDY OF THE ELECTRICAL POWER SYSTEM ON SYSTEM SAFETY ASSESSMENT

8.5.1 System Safety Assessment Inputs

The SSA of EPS takes the failure conditions determined in Chapter 4, System Functional Hazard Assessment as the safety objectives and combines them with the CCA results of relevant cases in Chapter 6, Common Cause Analysis. This analysis also assumes that the system architecture and interface described in Chapter 5, Preliminary System Safety Assessment, has been the stable version formed after several rounds of redesign and the PSSA iterative analysis, which establish the objectives and basis of SSA. Meanwhile, the requirements and assumptions derived from the PSSA are also to be validated and verified during the SSA process.

8.5.2 Verification of Top-Level Safety Requirements of Failure Conditions in System Functional Hazard Assessment

8.5.2.1 Confirmation of Validating the Classification of Failure Conditions

Assume that the authority and applicant have different opinions on the classification of the failure conditions 24-FC-2 "Loss of AC Normal Network." The applicant thinks it is major, but the authority considers it as hazardous. In this case, the applicant is usually required to justify their opinion by analysis and flight tests.

8.5.2.2 Verification of the Safety Requirements of Failure Conditions in System Functional Hazard Assessment

The quantitative FTA should be done for the catastrophic, hazardous, and major failure conditions in SFHA in order to verify compliance with the safety objectives. In this case, the reliability data of RAT GCU determined in Chapter 7, Failure Modes and Effects Analysis, is used to calculate results, combining with reliability data of other specific components. The calculated results are as follows (excerpt)(Table 8.1):

From these results, we can see that the FTA quantitative calculated results of all failure conditions are less than the values of the safety requirements, which means that the design is in compliance with these objectives. The column "Supporting materials" shows the traceability to the validation documents for determining the classification of the failure conditions.

8.5.3 Verification of the Safety-Related Requirements from Preliminary System Safety Assessment

The safety-related requirements and assumptions derived from the PSSA case in Chapter 5, Preliminary System Safety Assessment, are validated and verified as follows:

8.5.3.1 Verification of the Derived Safety-Related Requirements

The validation activities of assumptions in PSSA are as follows (Table 8.2).

Table 8.1 FTA quantitative calculated results for failure conditions (excerpt)

FC	FC REF.	Classification	Safety objective	Verification method	Supporting material	FTA results
Total Loss of AC Network	24-FC-1	I	1.0E−9/FH	Qualitative and quantitative FMEA, Qualitative and quantitative FTA, CCA		2.562E−11
Loss of AC Normal Network	24-FC-2	III	1.0E−5/FH	Qualitative and quantitative FMEA, Qualitative and quantitative FTA, CCA		7.613E−09
Total Loss of DC Network	24-FC-3	I	1.0E−9/FH	Qualitative and quantitative FMEA, Qualitative and quantitative FTA, CCA		3.052E−11
Loss of DC Normal Network	24-FC-4	III	1.0 E−5/FH	Qualitative and quantitative FMEA, Qualitative and quantitative FTA		7.852E−09
…	…	…	…	…	…	

Table 8.2 Validation activities of assumptions in PSSA

REF	Descriptions of assumptions	Validation methods	Validation materials
24–ASS-1	The average flight duraiton of aircraft is 1.2 flight hours	Function and performance test	
24–ASS-2	The system can supply adequate power to the load under normal or emergency situations	Function and performance test	
24–ASS-3	The failure rate of components conforms to the exponential distribution	Reliability test	
24–ASS-4	The pilot can correctly utilize the HMI components relevant to the power system	Human factor test	
24–ASS-5	The monitor can provide the failure detection of 100% effective coverage to the components of executive function	Coverage analysis	
24–ASS-6	The regular functional checking intervals of INV: 2000 flight hours	MSG-3 analysis, and maintenance program inspection	
24–ASS-7	The regular checking intervals of deployment function and power supply of RAT system: 2000 flight hours	Function test MSG-3 analysis, and maintenance program inspection	

The verification of safety-related requirements derived from the PSSA process is as follows (Table 8.3):

8.5.3.2 Confirmation of the Maintenance Means for Hidden Failures and Candidate Certification Maintenance Requirement Items

After analyzing the FTA of catastrophic and hazardous failure conditions, the relevant hidden failures can be determined. The maintenance task requirements are provided and documented in the CCMR report.

In this case, we only list the hidden failures related to the failure condition 24-FC-3.

Table 8.3 Verification of safety-related requirements derived from the PSSA

REF	The description of derived requirements	Validation methods	Validation materials
24-DSR-1	Main battery and APU battery should discharge continually for at least 30 minutes	Function and performance test	
24-DSR-2	Under emergency conditions, L DC BUS should be isolated from L ESS DC BUS	CCA	
24-DSR-3	The LIDG/RIDG is independent with the RAT;	CCA	
24-DSR-4	The LIDG/RIDG is independent with the RAT manual release cable (considered from the aspects of design, installation, specific risks, etc.)	CCA	
24-DSR-5	The LIDG/RIDG is independent with the Weight-on-Wheel Signal interface with the landing system	CCA	
24-DSR-6	The LIDG/RIDG is independent with the avionics built-in testing signal	CCA	
24-DSR-7	The LIDG, RGCU, and RAT GCU are independent	CCA	
24-DSR-8	The LIDG, RIDG N2 speed interface, and AC ESS BUS are independent	CCA	

Table 8.4 shows the hidden failures in the fault trees of the failure condition 24-FC-3 and the failure rates. The corresponding risk exposure time can be obtained by the following methods:

The RAT system is the alternate power used only in an emergency situation, so that a RAT system failure cannot be accurately detected when the power system works normally. Although part of the RAT system components will conduct the functional test and built-in test when the aircraft starts the power system, some of the components still need to be analyzed with MSG-3 to obtain 2000 flight hours or the operational test interval of 12 months.

Taking 2000 flight hours as the exposure time to be calculated quantitatively in the failure condition assessment will meet the safety requirement of various failure conditions, so 2000 flight hours can be regarded as the exposure time of relevant RAT components.

After verification, it can be determined that the above hidden failures are documented in the CCMR analysis report of the aircraft power system (see Table 8.5) and are finally transformed into the CMR requirements.

Table 8.4 Hidden failures list in catastrophic failure conditions

REF	Hidden failures	Failure rates	Exposure time/ Flight Hour	Relevant failure conditions	Classification
EV13-RATIDGF	RAT mechanism/ generator failures	8.4E−06	2000	24-FC-3	I
EV13-UPLOCKF	Mechanical up lock failure	2.04E−06	2000	24-FC-3	I
EV13-ACTUATORF	Deployment actuator failure	1.52E−06	2000	24-FC-3	I
EV16-MANRELEF	Manual release cable failure	4E−07	2000	24-FC-3	I
EV13-ADCUFRAT	ADCU failure causes the dysfunction of RAT	1.64E−06	2000	24-FC-3	I
EV16-LOCKRELEF	Release solenoid failure (up lock)	1.57E−06	2000	24-FC-3	I
EV13-PUMPF	Restow pump failure	2.1E−09	2000	24-FC-3	I

8.5.3.3 Confirmation of the Compliance with the Development Assurance Level Requirements

• Compliance with the FDAL

According to the FDAL assigned in the PSSA, combining with the requirements of SAE ARP4754A [2], the compliance substantiation should be indexed as follows (Table 8.6):

• Compliance with the IDAL

According to the IDAL assigned to GCU, RAT GCU, BPCU, battery controller, and ADCU in the PSSA, combining with the requirements of RTCA DO-254 and DO-178C, the compliance substantiation should be indexed as follows (Table 8.7−8.9):

8.5.3.4 Confirmation of the Operational Procedure and Limitations Derived From the Safety Assessment

The above requirements of operational procedures derived from the abnormal situation and emergency are included in the AFM.

8.5.3.5 Correctness Checks for Master Minimum Equipment List

Assume that the MMEL of the aircraft's EPS is shown in Table 8.10.

Table 8.5 EPS CCMR program

REF	Hidden failure	Failure effects (final effect)	Cause of failure	Checking task	Task interval
24-CCMR–01	RAT fails to be deployed	Aircraft: In the emergency situation of power, the RAT cannot be deployed and the aircraft loses the AC	Deploy Solenoid failure (including electrical and mechanical failures)	The automatic deployment check of RAT system	2000FH/12 MO
			ADCU failure		
			The failures of release handle and cable assemblies	The manual deployment test of RAT system	
				The general visual inspection of RAT components	
			Failure of deployment actuator	The manual deployment test of RAT system	
			Mechanical failure of uplock		
			RAT fails to pivot around the axis	RAT components plus the lubricants	
24–CCMR–02	RAT generator failure	Aircraft: In the emergency situation of power, the RAT cannot be deployed and the aircraft loses the three–phrase AC	The failure of RAT components	The power supply test of RAT system	

Note: Task interval time: 2000FH/12 MO refers to 2000 flight hours or 12 months; the first achieved prevails.

Table 8.6 Verification of EPS Development According to FDAL

REF	Function	FDAL	Subfunction	FDAL	Compliance materials
1	AC Power Supply	A	Normal AC Power Supply	B	The EPS CP The EPS Configuration Index (CI) EPS Certification Summary (CS)
2			Emergency AC Power Supply	B	
3			Independence	A	
4	DC Power Supply	A	Normal DC Power Supply	B	
5			Emergency DC Power Supply	B	
6			Independence	A	

Table 8.7 Verification of hardware in EPS according to DAL

REF	Equipment	IDAL	Compliance materials
1	GCU	B	The PHAC on the GCU/RAT GCU/BPCU hardware
2	RAT GCU	B	The HAS on the GCU/RAT GCU/BPCU hardware
3	BPCU	C	The HCI on the GCU/RAT GCU/BPCU hardware
4	Battery control	B	The PHAC on the battery controller hardware The HAS on the battery controller hardware The HCI on the battery controller hardware
5	ADCU	B	The PHAC on the ADCU hardware The HAS on the ADCU hardware The HCI on the ADCU hardware

Table 8.8 Verification of software in EPS according to the DAL

REF	Equipment	IDAL	Compliance materials
1	GCU	B	The PSAC on the GCU/BPCU/ADCU software
2	BPCU	C	The SAS on the GCU/BPCU/ADCU software
3	ADCU	B	The SCI on the GCU/BPCU/ADCU software

The failure of LTRU will cause the temporary failure of LDC BUS, while the failure of RTRU will cause the temporary failure of RDC BUS. However, the LDC and RDC can back up for each other, and if one of the DC BUSs fails, the other can continually supply power to it by TIE BUS without causing catastrophic, hazardous, and

Table 8.9 Operational procedure requirements derived from the safety analysis in the abnormal and emergency situations

Derived requirements REF	Failure conditions	Derived requirements		Sources of requirements	Relevant explanations
		Flight crew instruction	Flight crew operation		
24-AFM-01	The failure of RAT automatic deployment	EICAS: BATT DISCHARGING LAMPS: None	Pull the manual deployment handle	24-FC-3 Total Loss of DC Network	When the automatic deployment fails, the crew should be warned to manually deploy the functions; otherwise it will cause catastrophic effects.
24-AFM-02	Generator failure	EICAS: LGEN OFF, RGEN OFF, or APU GEN OFF. In addition, it may present such information as LIDG OIL RIDG OIL or LOAD SHED. LAMPS: LGEN OFF, RGEN OFF or APU GEN OFF, LIDG FAULT, RIDG FAULT, and other relevant indications.	Try to restart the generator; if this fails, turn the switch to the OFF position.	24-FC-3 Total Loss of DC Network	During a generator failure, the crew should be warned to close the generator to prevent it from affecting the safety of the aircraft. It should also prevent the false protection of the generator, which will cause catastrophic failure conditions.
24-AFM-03	BUS failure: main AC, DC, ESS AC, ESS DC, conversion and flight control	EICAS: LDC ESS BUS, AC ESS1 BU, AC ESS3 BUS, DC ESS TRANS, FC BUS, LAC BUS, L DC BUS, R AC BUS, R DC BUS, or DC ESS BUS and other relevant information. LAMPS: L BUS TIE ISLN, R BUS TIE ISLN, DC BUS TIE ISLN, ETRU FAULT, L GEN OFF, R GEN OFF, LIDG FAULT, R	For LDC ESS BUS or RDC ESS, the DC BUS Tie should be reset. For DC ESS TRANS, the ETRU switch and EC TIE BUS should be reset; the failure of flight control BUS results in a nondispatch condition. The remaining indications will also result	24-FC-3 Total Loss of DC Network	During a BUS failure, the crew should be warned to disconnect the corresponding BUS to prevent it from affecting the safety of the aircraft.

(*Continued*)

Table 8.9 (Continued)

Derived requirements REF	Failure conditions	Flight crew instruction	Flight crew operation	Sources of requirements	Relevant explanations
		IDG FAULT or APU GEN OFF and other relevant information.	in a nondispatch condition.		
24-AFM-04	Storage battery overheat	EICAS: MAIN/APU BATT TEMP or MAIN/APU BATT OVERHEAT information LAMPS: None	As for MAIN/APU BATT TEMP and MAIN/APU BATT OVERHEAT, turn the switch of MAIN BATT or APU BATT to the OFF position.	24-FC-3 Total Loss of DC Network	In emergency cases, the storage battery should supply power to the deployment process of RAT, so that the abnormal situation should be checked to make sure RAT works normally.
24-AFM-05	Storage battery discharge	EICAS: MIAN/APU BATT DISCHARGE, BATT DISCHARGING information LAMPS: None	As for MAIN/APU BATT DISCHARGE and BATT DISCHARGIN, reset the switch of ETRU.	24-FC-3 Total Loss of DC Network	In emergency cases, the storage battery should supply power to the deployment process of RAT, so that the abnormal situation should be checked to make sure RAT works normally.

Table 8.10 EPS MMEL

EPS MMEL	Quantity	Dispatch quantity	Remarks
Main battery and APU battery	2	2	
RAT system	1	1	
ETRU	1	1	
ETRUC	1	1	
LTRU, RTRU	2	1	(LDC, RDC) DC BUS: The aircraft can be dispatched with the failure of one DC BUS. Besides, "RIDG, RGCU, RTRU, RTRUC, RESSC" or "LIDG, LGCU, LTRU, LTRUC, LESSC" must work normally.
LTRUC, RTRUC	2	1	(LDC, RDC) DC BUS: The aircraft can be dispatched with the failure of one DC BUS. In addition, "RIDG, RGCU, RTRU, RTRUC, RESSC" or "LIDG, LGCU, LTRU, LTRUC, LESSC" must work normally.
LESSC, RESSC	2	1	(LDC ESS, RDC ESS) DC ESS BUS: If DC ESS transfer BUS work normally, then the aircraft can be dispatched with the failure of one DC ESS BUS. Besides, "RIDG, RGCU, RTRU, RTRUC, RESSC" or "LIDG, LGCU, LTRU, LTRUC, LESSC" must work normally.
LIDG, RIDG, APU GEN	3	2	(LAC, RAC) Main AC BUS: If the APU generator and AGCU as well as operating IDG and GCU work normally, then the aircraft can be dispatched when the other IDG or GCU fails to operate. However, the AGCU and GCU must have the same design. (LAC, RAC) Main AC TIE BUS: If the RIDG and RGCU as well as the operating LIDG and LGCU work normally, then the aircraft can be dispatched when the APU or AGCU fails to operate.
RGCU, LGCU, AGCU	3	2	(LAC, RAC) Main AC BUS: If the APU generator and AGCU as well as the operating IDG and GCU work normally, then the aircraft can be dispatched when the other IDG or GCU fails to operate. However, the AGCU and GCU must have the same design. (LAC, RAC) Main AC TIE BUS: If the RIDG and RGCU as well as the operating LIDG and LGCU work normally, then the aircraft can be dispatched when the APU or AGCU fails to operate.

major failure conditions. However, in order to ensure the normal working of the other DC BUS, "RIDG, RGCU, RTRU, RTRUC, RESSC" or "LIDG, LGCU, LTRU, LTRUC, LESSC" must work normally at this time. Therefore, if one of the LTRU and RTRU fails, but the relevant requirements being satisfied, the aircraft can be dispatched, which still meet airworthiness safety requirements for MMEL.

Similarly, if one of the LTRUC and RTRUC fails, but the relevant requirements being satisfied, the aircraft can be dispatched, which still meet airworthiness safety requirements for MMEL.

The failures of LESSC and RESSC will, respectively, cause the temporary failure of LDC ESS and RDC ESS. However, if the LDC ESS and RDC ESS can back up for each other, then if even one of DC BUSs fails, the other can continually supply power to it by BUS TIE without causing catastrophic, hazardous and major failure conditions. However, in order to ensure the normal working of the other DC ESS BUS, "RIDG, RGCU, RTRU, RTRUC, RESSC" or "LIDG, LGCU, LTRU, LTRUC, LESSC" must work normally at this time. Therefore, if one of the LESSC and RESSC fails, but the relevant requirements being satisfied, then the aircraft can be dispatched, which still meet airworthiness safety requirements for MMEL.

Since the backup of APU and AGCU are not considered in the safety assessment processes, and the aircraft can satisfy minimum safety requirements when they are failed, and the failures of APU, AGCU can be listed in the MMEL report.

8.5.4 Checking for the Traceability of Safety Data

At the end of the safety assessment process, the traceability of all data related to system safety will be checked according to the traceability model in Section 8.3.3.

REFERENCES

[1] SAE ARP 4761. Guidelines and methods for conducting the safety assessment process on civil airborne systems and equipment. Society Automotive Engineers; 1996.
[2] SAE ARP 4754A. Guidelines for development of civil aircraft and systems. Society Automotive Engineers; 2010.
[3] RTCA DO-178C. Software considerations in airborne systems and equipment certification; 2011.
[4] RTCA DO-254. Design assurance guidance for airborne electronic hardware; 2000.

CHAPTER 9

Single Event Effects in Avionics

Contents

9.1 SINGLE EVENT EFFECTS

9.1.1 What Is a Single Event Effect?

Human beings live in an environment that is full of radiation that consists of all types of particles, including cosmic rays in space, particle radiation of the sun, and earth's electromagnetic radiation [1]. In the radiation environment, an integrated circuit could be affected, resulting in a Single Event Effect (SEE) [2]. A SEE refers to the injection of high-energy particles (e.g., high-energy protons and alpha particles) into the PN junction of integrated circuits, leading to the following: PN junction ionization that produces energy transfer and charge accumulation, causing circuit state mutations.

Civil Aircraft Electrical Power System Safety Assessment
DOI: http://dx.doi.org/10.1016/B978-0-08-100721-1.00009-1
239

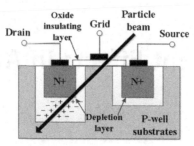

Figure 9.1 Single-event upset caused by heavy ions.

When a high-energy particle passes through the zone of the PN junction, a portion of the particle energy is absorbed by the silicon atom in its path. In addition, the particle creates electron–hole pairs in silicon. These electron–hole pairs undergo drift and diffusion movements under the electric field of the PN junction, thereby inducing a few tenths of nanoseconds of impulse current after the charge is collected. The transient current can change the potential of the node and, to a certain extent, can turn the conducting tube off and the blanking tube on, causing the logic state of the device to rollover; such a phenomenon is called a Single Event Upset (SEU) [3,4,5]. As shown in Fig. 9.1, heavy ions generate electron–hole pairs through the PN junction.

The SEU index is described by the SEU rate, which is based on the probability of the occurrence of SEU in devices each day at each bit. The general formula for calculating the proton SEU rate is as follows:

$$R_P = \int_{E_0}^{\infty} \sigma_P(E)\varphi(E)dE \qquad (9.1)$$

In formula (9.1), R_P is the proton SEU rate, whose unit is SEU/bit/d; E_0 is threshold energy, and its unit is MeV; $\sigma_P(E)$ is the proton SEU cross section, and its unit is cm^2/bit; and $\varphi(E)$ is the proton differential flow.

SEE includes destructive effect and nondestructive effect.

1. Destructive effect

   For example, Single Event Latchup, Single Event Snapback, Single Event Dielectric Rupture, Single Event Gate Rupture, and Single Event Burnout.

2. Nondestructive effect

   For example, Single Event Transient, Single Event Disturb, SEU, Multiple Bit Upset, and Single Event Functional Interrupt.

9.1.2 Single Event Effect of Static Random Access Memory Field Programmable Gate Array

SEE is a serious space radiation effect that causes airborne electronic equipment failure. The interaction of the particles and PN junction can lead the device storage unit to become a "bit upset", which is referred to as a "soft error" [6]. Such a phenomenon, which is induced by the radiation effect, can be rewritten or repaired by reset without permanent damage for a semiconductor device. These soft errors are mainly caused by SEU [7] and SET, with a particularly serious impact on Static Random Access Memory (SRAM) Field Programmable Gate Array (FPGA) circuits [8,9,10]. Therefore, these effects have attracted much attention. Among these effects, SEU is a charged radiation effect that occurs in monostable or bistable logic devices and logic circuits.

During the flight of an aircraft, complex and changeable external environment factors can cause some degree of impact on the airborne electronic equipment, seriously affecting the performance of the aircraft flight. Because SRAM-type storage chips have high density, low cost, reconfiguration, and other major advantages, they are becoming more and more popular in the field of aviation. However, SRAM-type devices are more susceptible to SEEs, leading to serious safety issues during aviation applications.

At present, mainstream FPGA technology consists of two different major technologies: one is based on the anti-fuse and the other is based on SRAM. The traditional space device is based on the anti-fuse type of FPGA because the storage structure based on anti-fuse will not lose data when the power supply drops after FPGA programmed. These FPGAs are convenient, stable, reliable, and have a superior resistance to radiation performance. However, with the development of space technology, the demand on signal processing ability is increasing and the development cycle of the task is decreasing [11], thereby placing higher requirements on the processor and integrated circuit. However, the FPGA that is based on anti-fuse cannot satisfy these demands [12]. To satisfy the demands of greatly reducing costs, improving equipment performance, and shortening the development cycle, we can use commercial grade devices in airborne electronic equipment; such devices are called COTS devices.

SRAM, as a typical representative of COTS devices, is more sensitive to space radiation; however, its performance is good, it has a low cost, its supply is adequate, and it can be repeatedly programmed. In contrast, although the FPGA based on anti-fuse has strong radiation resistance, its cost is high, its production cycle is long, and it has poor performance. The FPGA that is based on SRAM has better commercial advantages than the FPGA based on anti-fuse; as a result, SRAM FPGA is increasingly used in the field of aviation and space flight.

9.2 SINGLE EVENT EFFECT FAULT TEST METHOD—HOW TO TEST?

Based on the research of the SRAM FPGA configuration memory structure and configuration of the structure, the researchers found that the information of the FPGA configuration file 0/1, which can represent the upset, could be used to simulate the SEU effect of FPGA through changing the configuration storage. It provide the theoretical foundation of SRAM FPGA fault injection.

The current approaches commonly used for evaluating the upset effect of FPGA include: the flight experiment [13,14,15], ground radiation test [16,17], and fault simulation technology.

9.2.1 Flight Experiments

The flight experiment aimed to directly test the target device in a flight environment. The advantage of this experiment is obvious, i.e., carrying out a test in a real radiation environment allows the most primitive data to be collected and is the simplest SEE reverse evaluation method. However, the cost of the satellites carrying the experiment is very high, and the test cycle is long, not artificially controlled during the experiment, and the test lacks flexibility.

The Xilinx Company produces a FPGA, Xilinx Virtex-II 2V6000, with the number of system gates of 6 M. The FPGA is manufactured with 150 nm fabrication processes. In terms of the Xilinx FPGA, the Configurable Logic Blocks (CLBs) and Look-Up-Tables (LUTs) and the Block Random Access Memory (BRAM) are sensitive to the SEU. In addition, the Xilinx Company has performed a series of studies on SEU on this chip [18].

Flight experiments used the Rosetta experiment, each group of which contained 100 Xilinx FPGAs. Experiments were conducted at four different altitudes, and all of the test components were produced by United Microelectronics Corporation (UMC, United Microelectronics Corporation). Fig. 9.2 shows the testing device of 100 Xilinx FPGAs.

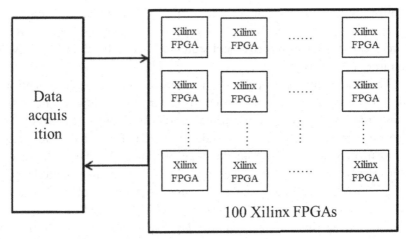

Figure 9.2 One test set of 100 Xilinx FPGAs [18].

Flight experiments at different altitudes mainly studied particle radiation effects on Xilinx FPGA. In addition, atmospheric experiments were performed at four sites: in San Jose in the United States, in New Mexico in the United States, in the White Mountains, and in Mauna Kea. The four elevations were 0, 5100, 12,470 and 13,200 ft [19,20]. At different flight altitudes, it was shown that 350 SEUs were detected in the long flight experiment data of 7050 devices–years.

9.2.2 Ground Test Equipment

Ground radiation testing refers to the ground on which a particle accelerator offers all types of particle sources to simulate the space environment to test the target device [21]; this approach is by far the most commonly used evaluation method. Each year in the United States, NASA selects devices in this manner, publishes results and provides references for researchers. For several large manufacturers, including Xilinx and Actel, radiation tests are important to their products and provide radiation reports. The ground radiation test data are real, but their disadvantages include that it requires the expensive equipment, the test chip suffers damage, and the experiment is difficult to perform.

9.2.2.1 Pu-Be Neutron Source Test System

For the operation altitude of civil aircraft, the primary radiation particles in the atmosphere are neutrons. Therefore, it is necessary to perform neutron radiation experiments [22]. For ground radiation simulation experiments, the Pu–Be neutron source is very appropriate because it is very easy to obtain, process, and store and it has the ability to provide continuous emission.

The test system has a Pu–Be neutron source and can produce neutrons. The Pu–Be neutron source (a 3.8-cm diameter, 9.1-cm long cylindrical) is transferred to the neutron pit where it is placed on a test stand. The test card contains the devices under test (DUTs), which are placed approximately 1 cm away from the edge of the source. Through correlation calculation, the neutron flux at the DUT location is approximately 2×10^5 n/cm^{2}s. The computer system used above includes the following: CPU, 2 M memory, RS232 connection cable, test card, analog input/output, and 64 threads input/output. The computer system is located in the pit at one end of the room; the room can shield the effects of neutrons by neutron-shielding. Only the test card connecting cable and part of the cable in the neutron source can be subjected to neutron radiation.

9.2.2.2 The Cyclotron of Louvain la Neuve Test Device in the European Space Agency

The European Space Agency used the particle beam produced by the Cyclotron of Louvain la Neuve (CYCLONE) [23]. CYCLONE can produce multiple particles, variable energies of accelerated protons (energy reaches 75 MeV), alpha particles, and heavy ions. For heavy ions, the energy range is from 0.6 MeV to 27.5 MeV. Heavy ions in bipolar electron cyclotron resonance (ECR, electron cyclotron resonance) are produced.

The heavy ion radiation test device, as shown in Fig. 9.3, consists of a test chamber and a beam monitoring system. The test chamber has a cylinder with a diameter of 50 cm and a length of 50 cm and is fitted with a 25 cm × 25 cm framework. Ion transmission allows the use of a multiple device communication board and rotation (−70 degree to 90 degree) to obtain the value of a variable effective LET. Based on the

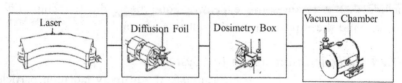

Figure 9.3 Test device for heavy ion radiation [22].

measurement, the device can have an effective LET range between 2.97 MeV/mg/cm^2 and 111.8 MeV/mg/cm^2.

To obtain a good homogeneous beam, a specification diffusion foil with 1.53 mg/cm^2 gold is placed in the front of the test chamber. Using this device, a beam that is 20 mm in diameter is obtained that is uniform over $\pm$ 10%.

The ion flux and purity can be controled by a surface barrier detector. The detector is placed in the test chamber and connects with a multichannel analyzer (MCA, Multi-channel Analyzer) and a speed meter. During the radiation, beam flux is monitored by using a parallel plate avalanche counter (PPAC, Parallel Plate Avalanche Counter) which is placed in the front of the device; the PPAC detector is a transmission chamber located below the gas circulation and is used to measure the total flux. The ion flux maximum available is 1E4 particles/s/cm^2, and the limit is decided by the Dosimetry System.

9.2.3 Fault Simulation Technology

Compared with these two methods, which are complicated and have a high cost described above as well as low flexibility, fault simulation technology is used to establish a model to test the target device in university laboratory and research institutes and has been used to study as well. According to the technical characteristics, the general components include the failure simulation technology, static failure analysis technology, and fault injection technique.

Fault simulation technology is more often applied than other technologies, and it is typically used in a simple system to build a simulation model under different conditions to study the different features of the fault and different hardware. Hardware description language is used to establish a model of a process that is more complex and poorer operability. Static failure analysis technology is based on establishing the necessary model to analyze the sensitivity of the FPGA designed to SEE; this method is often used because of its high accuracy. The early fault injection technique [24,25] was proposed in the 1970s and had been applied to the validation of fault-tolerant systems. Subsequently, it was introduced into the IC field, and after being introduced into a circuit, the output observation and analyses in fault cases of output are obtained. Finally, quantitative or qualitative reliability evaluation results are obtained; this approach is by far the most common fault simulation method.

Several assessments of the above technology are applicable to SRAM FPGA. However, because of the ground radiation experiment and satellite experiment, which have high costs and long testing times, only the national space agency, key laboratories, and some large companies can afford it. For researchers and research institutions, the low cost and flexible fault injection test has become the most important research object; such a test is heavily promoted. The FPGA configuration data can be changed after making selections by the fault model to achieve fault injection; fault injection controllability and observability are very high, while their cost is low.

9.3 SEE FAILURE FOR ELECTRONIC DEVICES IN AVIONICS

9.3.1 130-nm Xilinx Process Dimension Field Programmable Gate Array Flight Experiment

Virtex-II Pro, produced by Xilinx Company, is an SRAM FPGA with a 130-nm production process. The researchers [18] of Xilinx selected 200 chips with the model of XC2VP50 FPGA to conduct atmospheric radiation experiments; the experimental atmospheric altitudes were 0, 5100, and 12,470 ft. The experiment is based on the JESD89 standard. The test devices used by the research institutions this time are shown in Table 9.1. As a result of the atmospheric experiment, the failure rate was 290 FIT/Mb in the configuration unit and 530 FIT/Mb in the BRAM. The experimental results are based on the data statistics of 5900 devices and 317 SEUs phenomena.

Virtex-II Pro is an SRAM FPGA with a 130-nm production process. The failure rate in the configuration unit resulting from the atmospheric experiment was 290 FIT/Mb, and it was 530 FIT/Mb in the BRAM. In the Virtex-II which is an FPGA with a 150-nm production process, it was 295 FIT/Mb, and in the BRAM, it was 265 FIT/Mb. Obviously, the failure rate in the configuration unit of the two FPGAs are basically consistent, but the failure rate of Virtex-II Pro in the BRAM was twice that

Table 9.1 Upset experiments at four different altitudes for the XC2VP50 FPGA [18]

Device type	Technics	Altitude/ft	DUT numbers	Device hours
Xilinx Virtex-II Pro XC2VP50	130 nm	0	200	1,191,000
		5100	200	709,000
		12,470	200	65,500

of Virtex-II. The main reason for the difference is that the 2VP50 and 2V6000 configuration latches are in same design, and thus the failure rate in the configuration unit is basically consistent. However, because the process of 2VP50 in the BRAM is 130 nm and that of 2V6000 is 150 nm, the failure rate of 2VP50 in the BRAM is much higher than that of 2V6000.

9.3.2 Flight Experiment of a 90-nm Xilinx Process Dimension FPGA

Xilinx Virtex-4 FPGA is a typical FPGA chip with a high usage rate. Many research institutions have conducted relevant research on it; e.g., Sandia National Laboratory, NASA, and Xilinx have placed a virtex-4 device in a near-earth orbit to observe the upset failure of SEE.

Los Alamos National Laboratory [26] conducted an experimental research on a Xilinx virtex-4 FPGA in 2012. The experiment was carried out on a satellite of the Department of Defense. The test system adopted by the research institution is a Los Alamos Experimental Unit (LEU). Each LEU contains two virtex-4 FPGAs, and the two chips models are XQR4VLX200 and XQR4VSX55. The purpose of the experiment conducted by the research institution is to determine the reliability of these devices when they are used in equipment with high reliability.

Two LEUs were used in this experiment, LEU1 and LEU2. There were 4 FPGAs; 11,330 SEUs phenomena were found during measuring them and the statistics are shown in Table 9.2. The experimental results showed that the average SEU rate of the XQR4VLX200 FGPA was 14.4 ± 0.13 SEUs/device/day and the average SEU rate of the XQR4VSX55 FPGA was 5.2 ±0.13 SEUs/device/day.

It can be seen from the experimental statistics data that, after being affected by the high-energy particle radiation in space, the FPGA SEU rate is very high. The daily average reversal phenomenon of each XQR4VSX55 device is 5.2 SEUs/device/day. The number of XQR4VLX200 is higher, reaching 14.4 SEUs/device/day on average. For high-reliability applications, such a high SEU rate may easily cause system failures.

Table 9.2 The ratio of device-days between the LX and SX [26]

	SEUs/device-day	R.S.D.
LX FPGA	14.4	± 0.13
SX FPGA	5.2	± 0.13

9.3.3 65-nm Xilinx Process Dimension FPGA Flight Experiment

Virtex-5 Xilinx is an FPGA with a 65-nm production process, and its internal configuration bits reach a million stage.

1. Virtex-5 Xilinx aviation altitude flight experiment

   MICROSEMI Corporation [27] conducted an experiment on a Xilinx Virtex-5 FPGA at an altitude of 40,000 ft. The experiment was conducted on the route of Tokyo—New York. For example, if the upset rate per Mb of the flight relevant closely to two-stage is 18.63 FIT/Mb and the configuration memory is 29.68 Mb, then the neutron flux at the altitude of 40,000 ft is 561.7 times the neutron flux of the benchmark. Consequently, the upset rate of the device was 3.1E6 FIT/device, i.e., to say that the failure rate per Mb is 1.05E5 FIT/Mb. This means that assuming that there are 4—5 FPGAs per LRU and 20 LRUs per aircraft in total, and the average upset time of the aircraft is 3.5 hours.

2. Virtex-5 Xilinx ground simulation experiment

   Researchers at Los Alamos National Laboratory have conducted single-event effect [28] experiments on a Virtex-5 FPGA. A ground radiation experiment that adopted neutron source radiation was carried out at the Laboratory. The experimental results showed that the bit-flipping section of the configuration unit is $6.7E-15$ cm^2/bit and the failure rate is 165 FIT/Mb; the bit-flipping section of the BRAM is $3.96E-14$ cm^2/bit, and the failure rate is 692 FIT/Mb. From the experimental results, it can be seen that the failure rate of neutron radiation in the BRAM is several times higher than in the configuration unit and much higher than 150- and 90-nm products.

   According to the experimental results of various aspects, for an FPGA, especially an FPGA based on SRAM, the failure rate triggered by SEU is too high and has a serious impact on the flight safety of civil aircraft. Aviation authorities of the Europe and the United States have been aware of the problem and have issued airworthiness documents to advise the aviation bureau, the industry and other relevant departments to consider the issue.

   With the continuous improvement of technology, from 180, 150, 130, and 90 nm down to 65 and 40 nm, it can be speculated that the neutron cross sections and SEE failure rate of the BRAM will continue to deteriorate. In SRAM FPGAs, the upset phenomenon is likely to occur and cause system failures. Especially in those applications areas required by high reliability, such as aircraft, the corresponding reinforcement and protection are essential; with the increase of altitude and particle radiation energy, the SEE is also increasing significantly. The SEUs in SRAM FPGAs caused by particles radiation at military altitudes are much more than at commercial altitudes. Even at the normal flight altitude, if no corresponding radiation reinforcement measure is taken, the SEU effect is still unable to meet the current airworthiness requirements.

9.4 SAFETY CONSIDERATION FOR SINGLE EVENT EFFECT IN AVIONICS

9.4.1 Industrial Standard

1. RTCA/DO-254

RTCA/DO-254 [29], Design Assurance Guidance for Airborne Electronic Hardware. DO-254, was introduced in previous chapters and collected the best practices of airborne electronic hardware design and guarantee in the industry, advocating for key electronic equipment and other avionics systems that meet high safety requirements. The design and verification should take a top-down method. In addition, it is recommended that the consideration method of airborne electronic hardware on the SEE should follow the DO-254 requirements.

2. RTCA/DO-297

RTCA DO-297 [30], "Integrated Modular Avionics (IMA) Development Guidance and Certification Considerations" includes IMA design assurance—flexible, reusable, interoperable hardware and software to form a platform that allows multiple applications to operate on the same hardware. Regarding the fact of IMAs system, the SEE should be taken into consideration, only a few lines are mentioned.

9.4.2 FAA Requirements

1. AC 20-152

The FAA AC 20-152 [31], dated June 30, 2005, is specified that DO-254 is the official guidance material for suppliers of civil aviation avionics system. It is applicable to application specific integrated circuit (ASIC), programmable logic device (PLD), field programmable gate array (FPGA), or similar electronic components for the design of aircraft systems and equipment. AC indicated that DO-254 is applicable to assurance levels A to C of hardware design. The equipment at level D is free to choose whether to meet DO-254.

FAA defines many safety and importance levels of an avionics system, e.g., DAL A or B should conduct tests, verifications, and document processing strictly, whereas the requirements for DAL C, D, and E are much lower. The failure levels for all of the aircraft hardware are determined. In the actual application of airworthiness, systems with DAL A or B level require the consideration of adopting SEE reinforcement technology, whereas systems with lower levels do not need to consider it. For example, the safety of an airborne entertainment system will not be affected by the SEE. Engineers must analyze the reinforcement requirement of each system and the corresponding resource occupancy.

2. DOT/FAA/AR-95/31 Design, Test, and Certification Issues for Complex Integrated Circuits [32]

As early as 1993, a report by Keller indicated that system integrators, aviation electronic equipment manufacturers, and some major commercial aircraft manufacturers found that when an aircraft flew at an altitude of over 10,000−15,000 ft, problems caused by cosmic radiation may threaten flight safety.

3. FAA Final Report—Handbook for the Selection and Evaluation of Microprocessors for Airborne Systems

In February 2011, the FAA released the "Final Report—Handbook for the Selection and Evaluation of Microprocessors for Airborne Systems", a manual for the selection and evaluation of airborne system microprocessors [33]. This paper demonstrated, in detail, that airborne systems should consider the SEE when selecting a microprocessor.

4. AC 20-170

In October 2010, the FAA released AC20-170 [34], Development, Verification, Integration, and Approval of Integrated Modular Avionics based on RTCA DO-297 and TSO-C153, which officially recognized the RTCA DO-297 standard. In addition, it mentioned the issue of SEEs.

9.4.3 European Aviation Safety Agency Requirements

1. ED-80 EUROCAE

ED-80 [35] is the corresponding version of EUROCAE Europe. The document is consistent with DO-254 in content and applications.

2. EASA Safety Information Bulletin

Single Event Effects (SEEs) on Aircraft Systems caused by Cosmic Rays

EASA released a safety information bulletin titled "Single Event Effects (SEEs) on Aircraft Systems caused by Cosmic Rays" in

October 2012 [36]. The bulletin is used to inform aircraft operators, manufacturers, designers of aviation electronic systems, manufacturers of electronic equipment and components that SEEs may cause faults.

When individual particles (neutrons, atoms, or other heavy ions) that interact with carbons atom contained in a semiconductor device in the aviation system, a SEE may occur. Because the integrated circuit package for aircraft has diminished in size, the risk will be increased.

3. EASA Report

In the Safety Implications of the use of system-on-chip (SOC) on COTS devices in airborne critical applications [37] in on-chip system integration in aircraft application safety problems, released by the EASA in January 2008, the safety issue of SEU and multibit upsets are listed for consideration.

9.4.4 Safety Considerations for Single Event Effect

1. It is generally believed by the authorities and industry that, at the flight altitude, electronic devices are likely to affect aviation flight safety because of failures caused by cosmic radiation.
2. RTCA/DO-254 and EUROCAE/ED-80 are taken as the main standards for guiding the design of complex electronic hardware (CEH) in aviation systems, as approved by European and American Council. FAA AC 20-152, dated June 30, 2005, specified DO-254 as the official guidance material for suppliers of civil aviation avionics systems.
3. The FAA holds the opinion that aviation equipment of DAL A and B must consider the issue of SEEs, whereas equipment of DAL C may consider it or not.
4. Regarding avionics, the radiation effect shall be considered when selecting components to determine whether the device meets the requirements. The radiation sensitivity evaluation method can be implemented into an existing system development process; the main steps of this method include SEE sensitivity analysis, component selection and test, system analysis, and evaluation of reinforcement.

9.5 THE SENSITIVITY ASSESSMENT METHOD OF THE RADIATION EFFECTS

With the selection and adoption of sensitive devices, the use of these devices must meet the safety requirements as well. The sensitivity assessment method of the radiation effects can be integrated into system development and safety processes. The main steps include sensitivity analysis of SEE, component selection and test, system analysis and reinforcement evaluation, described as follows:

1. According to the DAL of the airborne electronic equipment under analysis, the decision whether to take SEE assessment and/or adopt additional design against SEE should be made. SEE problem must be solved on DAL A and B equipment. It is not required but suggested to consider SEE on DAL C. If there is no special requirement, it is not required for DAL D and E.

2. On the basis of the FFPA analysis, list the modules or devices which can be affected by SEE problem.

3. Identify sensitivity to SEE for each device in the list, according to the parameters such as fabrication technology and size. In general, memory, FPGAs, microprocessors, and those kinds of devices are sensitive and need further analysis.

4. Once the sensitive device is determined, detailed investigation and analysis should be performed for each individual device. Analyze whether the functions carried by the device are affected by the SEE. If not, record the analysis results, otherwise, continue to next step.

5. To integrate the device failure caused by SEE into safety process, the failure rate of this kind should be calculated. Therefore, the SEE radiation test data of device is needed for calculating. If the data cannot be obtained, the analyst may have to conduct an additional radiation test to get the data of target device. Normally, the radiation test data of one device cannot be used to evaluate other similar devices, since these devices may have different characteristics, unless sufficient proof can be provided.

6. According to the identified sensitivity for each device, the special failure rate should be calculated with consideration of both functional failure and physical failure caused by SEE (see NOTE in the end of section). The device failure caused by SEE may be taken as a failure mode into FMEA.

7. Compare failure rate with expected SEE failure rate within an order of magnitude. If there is a large gap, the reinforcement against SEE should be under the consideration. Additionally, if the failures caused by SEE have direct impact (or a kind of potential common mode) on system critical function, the reinforcement against SEE should be under consideration. If the reinforcement is needed, go to Step **8**. Otherwise, go to Step **9**.

8. Evaluate the feasibility of the reinforcement in technology and cost, etc., and make a decision of changing the design or employing reinforcement.
9. Adopt the SEE failure rate into safety assessment process and make detailed quantitative assessment.

NOTE: The calculation of the SEE failure rate in Step **6** may be very complicated, and it needs specific analysis and experiment with other technology, which is not in the scope of this book.

The safety assessment of SEU is an iterative process, and the assessment processes are shown in Fig. 9.4.

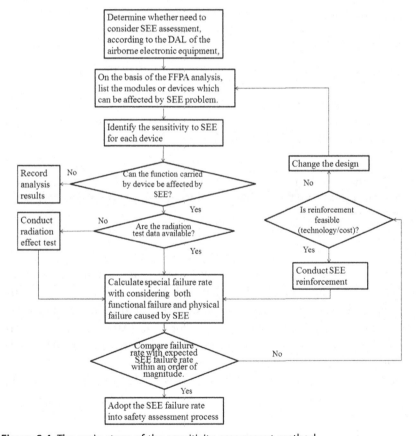

Figure 9.4 The main steps of the sensitivity assessment method.

9.6 THE EXAMPLE OF SINGLE EVENT EFFECT SENSITIVITY ASSESSMENT

Take the GCU unit of EPS as an example.

1. The GCU discussed is categorized as DAL B. So it needs to assess the SEE problem.
2. A determination is made that the modules in GCU need to carry out an SEE analysis.
3. A sensitive device is found. The FPGA chip X1 in GCU is sensitive.
4. The radiation test data for FPGA X1 is available from device OEM (Otherwise, the device will be sent to the radiation experiment laboratory to get the data.).
5. The failure rate (λ_{SEE}) is considered both with physical failure and functional failure of SEE.

   With the failure radiation test data, the SEE failure rate is calculated as follows:

   SEE rate per device hour = Integrated neutron flux ($n/cm^2/$ hour) $\times$ SEE cross-section ($cm^2/$device).

   The SEE cross-section parameter was got from the radiation test, including the experiment and functional test.

   In this case, the integrated neutron flux is 6000 $n/cm^2/$hour, and the SEE cross-section is 2E−10 $cm^2/$device. The SEE failure rate is calculated as 1.2E−6 per hour.

> NOTE: The parameter values used above are just applied to this case. They may be changed in different types of aircrafts due to different flight envelopes and chosen devices.

6. The allocated failure rate from the PSSA and preliminary FMEA is 1E − 6 per flight hour. The calculated λ_{SEE} is a little bit exceed the expected rate, but within the same order of magnitude. Therefore, reinforcement is not required at current stage (After all, the final conclusion depends on the result of the safety assessment, where the design met the safety requirements in this case.).
7. We take the failure caused by SEE as a special failure mode of this device (parallels with the original device physical failure mode, as a safety engineer known). We put these special failure rates into FMEA of RAT GCU (see Table 9.3).

Table 9.3 FMEA for the example

FMEA#	Failure mode/cause	Flight phase	Failure effect	Identification and corrective action of failures	Dispatch requirements with failures	The failure rate of component	Mode failure rate	Exposure time	Mode probability	Classification	Remarks
			1. Local effect 2. Higher level effect 3. Final effect (on aircraft)	1. Indication to the flight crew 2. Other failures with the same indication 3. Identification, isolation, and corrective actions of failures taken by the flight crew. 4. The effects of possible inappropriate actions 5. Failure isolation—maintenance personnel 6. Corrective action—maintenance personnel	1. If able, the aircraft can be dispatched. 2. If able, what is the flight limitation?	(1E-7/FH)	(1E-7/FH)	(FH)			
24-23-1-1.10	FPGA failure caused by single-event effect[1]	All	1. The RAT GCU excitation control on errors. 2. N/A 3. N/A	1. N/A 2. N/A 3. N/A 4. N/A 5. Reset RAT GCU in fight 6. Reset RAT GCU in fight	1. Able 2. After reset, it should pass BIT	Reference to note in this table	12	1.2	1.44E − 6	Hazardous	
24-23-1-1.11	Original random physical failure of FPGA	ALL	1. The RAT GCU excitation control was failed. 2. N/A 3. N/A	1. N/A 2. N/A 3. N/A 4. N/A 5. Replace RAT GCU 6. Replace RAT GCU	1. Disable 2. N/A	0.05	0.05	1.2	6E − 9	Hazardous	Obtained from the supplier

[1]This failure is only caused by SEE and has no relationship with FPGA quality, so there is only mode failure rate.

The following shows how to adopt the failure caused by SEE into the safety process by FMEA (Table 9.3(24-23-1-1.10)). The failure mode is the FPGA failure caused by single-event effects. The flight phase is "All." It is possible to make the RAT GCU excitation control on errors and has "Hazardous" failure effect.

The special failure mode (caused by SEE) of FPGA is different from the device physical failure mode of FPGA. Table 9.3(24-23-1-1.11**) also gives the physical failure for FPGA of RAT GCU as well. This "normal" failure rate of device was found from the device menu.

REFERENCES

[1] Ziegler JF, Curtis HW, Muhlfeld HP, et al. IBM experiments in soft fails in computer electronics (1978−1994). IBM J Res Dev 1996;40(1):3−18.

[2] Leray JL. Effects of atmospheric neutrons on devices at sea level and in avionics embedded systems. Microelectron Reliab 2007;47(9):1827−35.

[3] Xilinx, Single Event Upset Mitigation Selection Guide. America.

[4] Lawrence RK, Kelly AT. Single event effect induced multiple-cell upsets in a commercial 90 nm CMOS digital technology. IEEE Trans, Nucl Sci 2008; 55(6):3367−74.

[5] Binder D, Smith E, Holman A. Satellite anomalies from galactic cosmic ray. IEEE Trans Nucl Sci 1975;22(6):2675−80.

[6] Karnik T, Hazucha P, Patel J. Characterization of soft errors caused by single event upsets in CMOS processes. IEEE Trans Dependable Secure Comput 2004; 1(2):128−43.

[7] Zhu M, Song N, Pan X. Mitigation and experiment on neutron induced single-event upsets in SRAM-based FPGAs. IEEE Trans Nucl Sci 2013;60(4):3063−73.

[8] Wang T, Chen L, Dinh A, Bhuva B. Single-event transients effects on dynamic comparators in a 90 nm CMOS triple-well and dual-well technology. IEEE Trans Nucl Sci 2009;56(6):3556−60.

[9] Yang H, Sun J, Wang W. Review of the FPGA devices design technology development. J Electron Inform Technol 2010;32(3):714−27.

[10] Yan L, Wang Q, Fang L, et al. Application of the programmable logic devices in space electronics. Chinese J Space Sci 2009;29(1):54−8.

[11] Li H, Liu H, Yang G, et al. Study on astronautic single event effect fault injection system. Chinese J Quantum Electron 2002;19(1):57−60.

[12] Hillman R, Swift G, Layton P, et al. Space processor radiation mitigation and validation techniques for an 1,800 MIPS processor board. In: Radiation and its effects on components and systems (RADECS); 2003. p. 347−52.

[13] Chung SI, Lee HH. Analysis of the effects of high-energy electrons on a low earth orbit satellite parts. In: 6th IEEE international conference on industrial informatics (INDIN); 2008. p. 700−2.

[14] Harboe-Sorensen R, Poivey C, Fleurinck N, et al. The technology demonstration module on-board PROBA-II. IEEE Trans Nucl Sci 2011;58(3):1001−7.

[15] Caffrey M, Morgan K, Roussel-Dupre D, et al. On-orbit flight results from the reconfigurable Cibola Flight Experiment Satellite (CFESat). In: 17th IEEE symposium on field pro-grammable custom computing machines (FCCM); 2009. p. 3−10.

[16] Pace C, Libertino S, Crupi I, et al. Compact instrumentation for radiation tolerance test of flash memories in space environment. In: IEEE instrumentation and measurement technology conference (I2MTC); 2010. p. 652—5.

[17] Howe CL, Weller RA, Reed RA, et al. Role of heavy-ion nuclear reactions in determining on-orbit single event error rates. IEEE Trans Nucl Sci 2005; 52(6):2182—8.

[18] Lesea A, Drimer S, Fabula JJ, et al. The Rosetta experiment: atmospheric soft error rate testing in differing technology FPGAs. IEEE Trans Device Mater Reliab 2005; 5(3):317—28.

[19] White Mountain Research Station, San Diego, CA:University California, Office of Research. [Online]. Available: http://www.wmrs.edu/facilities/barcroft/barcroft. htm.

[20] Caltech Submillimeter Observatory, Hilo, HI. [Online]. Available: http://www.cso. caltech.edu/.

[21] Cannon EH, Cabanas-Holmen M, Wert J, Amort T, Brees R. Heavy ion, high-energy and low-energy proton SEE sensitivity of 90-nm RHBD SRAMs. IEEE Trans Nucl Sci 2010;57(6):3493—9.

[22] Normand E, Wert JL, Doherty WR. Use of Pu-Be source to simulate neutron-induced single event upsets in static RAMS. IEEE Trans Nucl Sci 1988; 35(6):1523—8.

[23] Berger G, Ryckewaert G, Harboe-Sorensen R, Adams L. The heavy ion irradiation facility at CYCLONE -a dedicated SEE beam line. In: IEEE radiation effects data workshop; 1996. p. 78—83.

[24] Portela-Garcia M, Lindoso A, Entrena L, et al. Using an FPGA-based fault injection technique to evaluate software robustness under SEEs: a case study. In: 12th latin american test workshop (LATW); 2011. p. 1—6.

[25] Wang X, Xu Z. A novel fault injection algorithm for safety analysis. In: Spring congress on engineering and technology (S-CET); 2012. p. 1—4.

[26] Quinn H, Graham P, Morgan K, et al. Flight experience of the Xilinx Virtex-4. IEEE Trans Nucl Sci 2013;60(4):2682—90.

[27] Microsemi. SoC products group. Understanding the impact of single event effects in avionics applications white paper.

[28] Quinn H, Morgan K, Graham P, et al. Static proton and heavy ion testing of the Xilinx Virtex-5 Device. IEEE Radiation Effects Data Workshop. New Jersey: Institute of Electrical and Electronics Engineers Inc; 2007.

[29] RTCA, DO-254: Design assurance guidance for airborne electronic hardware. RTCA, Inc., Washington, DC.

[30] RTCA DO-297:Integrated modular avionics (IMA) development guidance and certification considerations. RTCA, Inc., Washington, DC.

[31] Advisory Circular 20-152, RTCA, Inc. DOCUMENT RTCA/DO-254, design assurance guidance for airborne electronic hardware. Federal Aviation Administration, AIR-100, 2005.

[32] DOT/FAA/AR-95/31. Final report, design, test, and certification issues for complex integrated circuits. Office of Aviation Research Washington, D.C. 20591. 1996.

[33] FAA Final Report. Handbook for the selection and evaluation of microprocessors for airborne systems; 2011.

[34] Advisory Circular 20—170. Integrated modular avionics development, verification, integration, and approval using RTCA/DO-297 and technical standard order-C153; 2010.

[35] EUROCAE ED-80. Design assurance guidance for airborne electronic hardware; 2000.

[36] EASA Safety Information Bulletin. Single event effects (SEE) on airplane systems caused by cosmic rays, EASA SIB No: 2012—10.

[37] Research Project EASA. Safety implications of the use of system-on-chip (SOC) on commercial of-the-shelf (COTS) devices in airborne critical applications. 2008/1.

CHAPTER 10

Formal Model Based Safety Analysis Methods and the Application

Contents

10.1 THE REASON FOR APPLYING FORMAL MODEL BASED SAFETY ANALYSIS

The safety assessment method suggested by the SAE ARP4761 has been accepted by the civil aircraft airworthiness development, and its effectiveness has been proved in practice. However, the actual operational data indicates that some systems which have already carried out the Failure Modes and Effects Analysis (FMEA) and Fault Tree Analysis (FTA) still

Civil Aircraft Electrical Power System Safety Assessment
DOI: http://dx.doi.org/10.1016/B978-0-08-100721-1.00010-8

cannot satisfy the safety requirements of airworthiness standards. The main causes are the diversity of airborne system functions and the increasing integration and complexity of system components, in particular the massive use of software-intensive safety-critical systems; meanwhile, the traditional FMEA and FTA are mainly dependent on the personal skill and experience of safety analysts. Due to the limitation of individual cognitive competence, it is hard for them to deeply understand the analyzed systems and predict all the possible behaviors (including normal and abnormal behaviors). Therefore, these analysts may likely omit the system failure conditions or misjudge the effects of system failures, causing the loss of aircraft safety-critical functions, and further accidents or even catastrophic crashes. Besides, the traditional safety analysis processes are excessively dependent on the experience of experts, so it is hard to adopt this method in a complete and continuous manner without any error and respond timely to the effects of short-term behaviors (e.g., the design change) on the system. Therefore, it is urgently in need of method improvements so as to be applied to the increasingly complex airborne systems.

The formal methods are introduced into the airborne system safety analysis field by the academic and industrial communities. The safety analysis method, based on the formal model, utilizes the temporal logic or higher order predicate logic and other formal methods to conduct modeling, and to describe the system specifications to be verified. Then, the verification tools will be used to check whether the established system models satisfy the system specification requirements, and automatically analyze whether all the failure modes satisfy the system safety requirements. Compared with traditional methods, the formal model method is the way of automatically carrying out the verification and analysis processes with the aid of computer technologies, making more objective analyses and more reliable results. As for the system with the feature of complexity, the safety assessment method based on the formal model will better satisfy the design and analysis requirements.

10.2 THE MEANING OF FORMAL METHOD

Formal method refers to a method which analyzes and studies the formal structure of thinking in the science of logic. It is a scientific methodology of modern logic system, representing the creativity of modern logic thoughts. Besides, it is also a framework tool for constructing the modern logic theory, and an intermediary linking theoretical logic and specific logic. For the formal method, thinking forms are compared with different contents (mainly the proposition and reasoning) to find the connection methods among all parts. For example, the proposition contains connections of each concept, while the reasoning contains connections of various propositions. After these connections are found, the common structures are extracted from them, followed by introducing symbolic languages to express formal structures of proposition or reasoning through connections among symbols. Based on the mathematics, the formal method aims to establish an accurate and unambiguous semantics, which can effectively describe every phase of system development, thus making system structures meet the users' requirements with inherent rationality, correctness, and good maintainability.

Formal method concept has been applied since ancient times, and further developed and improved in the modern logic. This method has been widely applied in several fields, the mathematics, computer science, and artificial intelligence in particular. It can precisely reveal various logic law and formulate corresponding logic rules, thus making various theoretical systems more rigorous. At the same time, this method can also correctly train the thinking ability and improve its abstraction capability.

For the critical system which sets a high requirement for the safety and stability in every field, whether the realization of key properties in the software development satisfies the requirement, specification can be effectively controlled by strict logical derivation when adopting the formal method. For other software development methods, this can only be ensured by experience, many-time repeated refinements and inspections, resulting in a higher cost on software development compared with the formal method. Therefore, the formal method is of great significance in terms of improving software reliability and increasing the development efficiency of critical software.

10.2.1 Formal Specification

Formal specification, also known as formal description, is a mathematic description about what the system is supposed to do (i.e., the system functions are described by formal semantics such as mathematical symbol with precise semantic). It is the starting point of program design and program composition, and also serves as the foundation to verify the correctness and safety of the system. The described system includes behavioral characteristic, time characteristic, function characteristic, internal system structure, and so on. Among these characteristics, the typical one is that a description method based on the system state, which constructs state and behavioral characteristics of the system and other models by using the mathematic abstractions such as element, set, field, sequence, and mapping. System designers can use formal specification to examine whether the design document exists contradictions (i.e., consistency), whether it completely represents the object being described without any omission (i.e., completeness).

In conclusion, the formal specification can be defined as a process, on the basis of mathematic theory, abstracting the system in one certain or some levels, extracting system properties worthy of concern, and simulating the system behavior. According to different methods of formal specification, its languages can be divided into different categories: the algebraic language (e.g., OBJ, Clear, ASL, ACTOne/Two, etc.), process algebraic language (e.g., CSP [1], CCS, PVS, π calculus, etc.), and temporal logic language (e.g., PLTL, CTL [2], XYZ/E, UNITY, TLA, etc.). Based on different mathematic theories and specification methods, these specification languages varies considerably, while they have one common characteristic that each is composed of fundamental components and structural components. The former can be used to describe fundamental (atom) specification, while the later incorporates the fundamental components into a larger specification. The components serve as the key points in the study and design of formal specification and main basis of measuring the pros and cons of specification languages.

According to the characteristics of system described, the formal specification method can be divided into two categories. For the first method to describe the sequential system behavior, it is mainly represented by the mathematic structures such as set, relation and function, and the state transition is represented by the relation between antecedent and consequent (e.g., Z, B, VDM, Larch, etc.); for the second method to describe the concurrent system behavior, with the utilization of simple mathematical types (e.g., integer), every behavior is represented by the partial relation of sequence, tree and event (e.g., Petri net [3,4]), time automata, and so on. Usually, after conducting the formal specification for the system, the corresponding theories and methods are introduced to analyze or verify the property of system models.

10.2.2 Formal Verification

There is a close connection between formal verification and formal specification. Formal verification, based on the established specification, adopts the mathematics or logic to express the specification and property of the system. Besides, according to the reasoning of mathematical theories, the property of systems concerned can be proved to meet the expectations of specification. During the top-down development process, the formal verification strengthens designers' understanding on the system, helps find the design defects uneasy to be identified by other methods, and reveals the problems existing in the system design such as inconsistency, ambiguity, and incompleteness. As the viewpoint proposed by the *IEEE Trans Evol Comput* in 1996: with the constant improvement of design complexity, the formal verification will step into the production field from the laboratory, and the formal methods have proved its values and will enjoy broad prospects in the field of electronic design automation, the study of formal verification theory has increasingly attracted public attention.

Compared with other verification methods, the formal method has the following advantages: (1) the verification all incentive space, along with complete processes and conclusions; (2) the correctness of verification results is guaranteed by mathematical theories and independent of system incentive; (3) the expected output sequence can be generated without establishing the reference model in the verification process; (4) when errors are found in the verification, the error debugging information can be generated which is simple and easy understandable. In general, the formal verification methods can be divided into three categories: theorem proving, model checking, and equivalence checking.

1. Theorem Proving

    The objective of theorem proving is to prove the design correctness through formal logic, which is composed of axiom and reasoning rules. The logic architecture is used in the theory proving system to describe the design, and the lemma is used to describe a series of properties. It is required to prove the correctness of lemma through some reasoning rules. The logic can be divided into first-order logic and high-order logic, which can precisely express the system information to avoid the imprecision occurred in the process of describing system with natural languages.

    Theorem-proving system can deal with complex logic operations. The theorem-proving process depends on the axiom, reasoning rules, derived definitions, and middle lemma, and usually, some personnel or experts with certain theoretical knowledge are required to choose the reasoning routes and interactively accomplish the proving process. Many achievements have been made in the study of theorem proving, and some more mature theorem-proving systems include A Computational Logic for Applicative Common Lisp, Prototype Verification System (PVS), Higher Order Logic, and so on.

2. Model Checking

Sine E.M. Clarke et al. proposed the model checking method based on the sequential logic and finite state transition diagram in 1996, it has been deeply studied and extensively applied in various research institutes and laboratories at home and abroad with over 10 years of development. Compared with theorem-proving method, it has an advantage of higher-degree automation. The fundamental concept of model checking is to describe the sequential property of procedures or circuits by using the sequential logic, to express the behavior and structure of procedures or circuits by using the Kripke, and to verify whether it satisfies the temporal logic formula through traversing the Kripke structure.

At present, according to different algorithm used, the model checking methods can be divided into three categories: explicit model checking method, the model checking method based on the decision diagram, and the model checking method based on the satisfiability problem. Each of them has their own merits and demerits.

3. Equivalence Checking

The equivalence checking aims to realize the design transformation during the top-down design process, so as to exhaustively check whether different levels of the system functions keep consistent before and after.

According to the algorithm, the equivalence checking can be divided into two categories: symbol equivalence checking and incremental equivalence checking. The symbol equivalence checking is to express the problems as a specific set of symbols and to seek solutions by use of model-checking or theorem-proving method. The incremental equivalence checking, based on similarity of system structures under verification, judges the equivalence of local systems step by step and then achieves the equivalence of the whole system. At present, the theory and industrial application of equivalence checking have been relatively mature and its common tools include the MDG developed by the Montreal University, the Affirm developed by the Cadence Design Systems, Inc., the ESP-CV and Formality of Synopsys Inc., etc.

10.3 THE DEVELOPMENT OF FORMAL MODEL BASED SAFETY ANALYSIS

10.3.1 The Development of Formal Model Based Safety Analysis at System Level

Over the past decade, many automatic safety analysis methods have appeared, aiming at the safety-critical system with an increasing complexity. In addition, more and more airworthiness authorities or standard-setting organizations have come to realize that the automatic safety analyses should be accepted by the airworthiness certification as the technology basis. The Model-Based Safety Analysis (MBSA), the typical method of automatic analysis, will be explicitly prescribed as the safety engineering recommended practice for the aviation industry in the next revision of SAE ARP 4761. At present, the certification process has been performed in the flight control system of Dassault 7X aircraft with the use of models that are expressed by the AltaRica language-based data stream.

Since 2000, EU has initiated series of studies on the system safety analysis of the aircraft and has accomplished two projects successively: Enhanced Safety Assessment for Complex Systems (ESACS) [5] and Improvement of Safety Activities on Aeronautical Complex systems (ISAAC) [6]. Thereafter, EU has developed the project of More Integrated Systems Safety Assessment (MISSA) [7]. The primary objective of ESACS is the definition of a methodology, i.e., the compliance with design and safety assessment processes of industrial partners in this project. The support of tools, which can be integrated with that already used by their industrial partners, is also required for the methodology. The ESACS defines a series of critical steps and establishes a basic platform, the configuration of which varies with the use of different tools. Besides, the platform can share the same system structures and provide the same fundamental functions. The ISAAC draws on and expands the structure of ESACS and makes a deeper research in the improvement and comprehensiveness of safety analysis for complex airborne systems. This project aims at enhancing the capacity and efficiency of safety analysts and system designers so as to establish a model-based formal development technology which can consider the safety and reliability in the early stages of project. At present, the participant organizations of the accomplished MISSA project includes famous European aviation enterprises, such as Airbus, Dassault Aviation, Alenia Aeronautica, and research institutions and universities. With the application of the SAT-based model checking method, this project has analyzed the safety of high-lift system of A340, hydraulic pressure system of A320, and the aircraft fire protection system of Alenia Aeronautica [8−10].

The NASA developed the model-based safety assessment project in 2006, i.e., the simulation analysis on the wheel brake system safety described by the SAE ARP 4761 by applying the model-based formal principle [11]. Through adopting a model-based system development process, the system designers and safety analysts can share the same system model. They can automate the safety analysis by adding fault models and a certain proportion of controllable physical system, which can not only reduce the expenditure but also improve the quality of safety analysis. Besides, this project report also indicates that the model-based development activities mainly focus on the modeling of digital control system. This model can be used for multiple analyses such as model checking, completeness and consistency analysis, and so on. The model-based development tools usually contain the autocoders, which can generate code directly from models. In this report, the common commercial research tools used for supporting model-based development activities are also introduced, mainly including the Simulink, the Esterel, the SCADE, the Statemate of i-Logix, and the SpecTRM of Software Engineering, etc.

Furthermore, Meenakshi et al. used three open-source model checking tools—SPIN, NuSMV, and SAL, to verify the mode conversion logic of aircraft autopilot [12]. Elmqvist Jonas and Nadjm-Simin used the Prism, a probabilistic model checking tool, to calculate the high residual risk of Unmanned Aerial Vehicle [13]. Lindsay Peter et al. used the SAL, a symbol model checking tool, to perform the safety assessment on the hydraulic system of Airbus A320 [14]. Adriano Gomes et al. used the Prism to perform the quantitative safety analysis on the aircraft Actuator Control System [15].

10.3.2 The Application of Formal Model Based Safety Analysis in Hardware Design and Verification

With the increasing of hardware workload in the airborne equipment, whether it will completely meet the design requirement directly affects the aircraft safety. In order to ensure the aircraft safety and reliability, the hardware needs to be comprehensively verified before going into service, so as to ensure the safety of design. Differing from the application direction of the system's formal methods, the hardwares are mainly used to ensure the consistency between input and output in each design transformation in the design life cycle of the electronics hardware and are committed to finding the potential problems which are difficult to be verified by the traditional electronic verification methods.

During the design process of integrated circuits, modification may occur at different phase of design process. It is difficult to ensure its compliance with the initial design requirements if the netlist files are not verified. However, the platform testing used in the functional verification will take a very long time for the simulation for the netlist. Therefore, the industry adopts the formal methods to compare the equivalence between the netlist and reference models. This formal method is used in the chip verification process of the Rockwell Collins. The method is used in the process shown in Fig. 10.1 to ensure the equivalence between synthesized or modified netlist and RTL source code function by comparing that between input (reference unit) and output (module to be detected). Since it is not required to input vectors, much time for writing test cases has been saved. Furthermore, compared with the method of simulation, a more comprehensive verification result.

Formal methods are not only applied in IC design, but also play an indispensable role in solving the problems that are difficult to test in traditional verification, such as the timing issues caused by the clock domain crossing.

In the complex airborne electronic hardware design, the clock domain crossing signals will cause severe potential safety problems to the aircraft. This is tended to be ignored by using the present mainstream verification methods (including functional simulation, timing simulation, and static timing analysis). Some errors are found after several times of operations or long-time service, and will have a severe impact on the reliability and safety of airborne circuits. As for problems of multiclock domain, the formal methods can be used to conduct effective verification for the design to be detected. This method is a mathematical process of analyzing DUT design. The verification will prove the correctness of clock domain crossing assertion by illustrating all the data. However, if there is only one counter-example to prove the failure of the assertion, then the transmission error can be proved in the clock domain crossing signals.

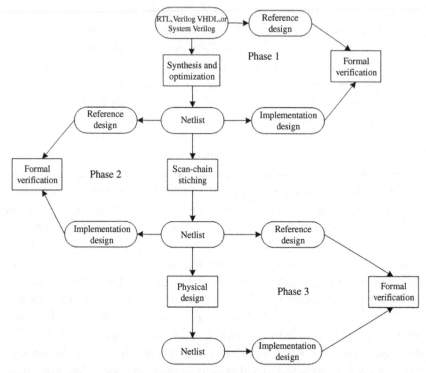

Figure 10.1 The application of formal verification in the IC design process.

At present, not only the formal method used in verification is indispensable, but also the function of formal method in the process of hardware requirement capture is gradually strengthened. The method used in requirement capture is model checking.

In the aeronautical field, it is often found in the validation of airborne electronic hardware that the logic of the design boundary to be tested is confused, the sequence is complex, the state transition is lost and so on. In most cases, the verifier is difficult to accurately locate these problems, which greatly extends the development cycle of the equipment, thus causing unnecessary time and economic losses. Therefore, it is necessary to use model checking method to ensure that the design is corretc and complete in time.

In conclusion, the formal verification for equivalence check can verify the functions of netlist in a more comprehensive and rapid manner and will ensure the uptime of design products. In addition, formal methods can help the designers get accurate design specifications or descriptions quickly. As for the problems of multiclock domain signal crossing which may severely affect the safety of large airborne electronic equipment, the formal verification technique has an irreplaceable advantage. The formal methods can effectively verify whether the airborne electronic hardware can meet its function and safety requirements.

10.3.3 The Application of Formal Model Based Safety Analysis in Software Design and Verification

With the increasing use of software in the aircraft systems and equipment, failure of airborne software will affect the safety of the system equipment and the overall aircraft. The application of formal methods in the verification process of airborne software will improve the performability, correctness, and safety dependency of the software, and will further ensure the safety of systems and equipment. Differing from the formal methods of systems, the formal methods are mainly used to check whether the software design meets its requirements.

The common software formal methods include B, Z, VDM, RAISE, etc. B is a formal method based on the mathematical theory and supports the overall development processes of software, from protocol description to producing the executable code. Z is a very popular formal specification language of software currently with the style of "state-operation." The method takes the first-order predicate logic and set theory as the basis of formal semantics and uses the function, mapping, correlation, and other mathematical methods for the specification. VDM is the abbreviation of Vienna Development Method, including a series of software formal specifications and software developing techniques. Derived from the research work of IBM Vienna laboratory, the VDM has a mathematical notation system and a proof rules based on the predicate logic and set theory. RAISE can also be called as the Rigorous Approach to Industrial Software Engineering. RAISE language is a formal language, and it can be used to write the extremely abstract and preliminary specification as well as the specific specification which can easily or even automatically convert into the programming language.

There are a series of comparatively mature tools for software formal verification methods, include SMV, Spin, UPPAL, CBMC, BLAST, SLAM, etc. SMV is a program used to verify the CTL formula by using the symbolic model algorithm. It adopts the finite state automata as the modeling language and describes the model and protocol in the input files. The model description language is a simple parallel assignment while using SMV. Spin is a system supporting the design and verification of asynchronous process, and it focuses on proving the correctness of process interaction. The process interaction can be realized through the synchronization primitive, asynchronous communication of pipelines, and shared variable. The UPPAL is a comprehensive tool platform to model, validate, and verify the real-time system. It models the real-time system as the timed automaton and extends its data types (including bound integer, array etc.). CBMC is a model checker which can verify the ANSI-C and C++ language, and it is used to verify the array bound, buffer overflow, pointer safety, exception, user defined assertion, and so on. BLAST is another model checker to verify the C language. It uses the predicate abstraction to abstract the C code and provide the automatic verification for program. SLAM is a formal verification tool based on the model checking and is used to check whether the software interfaces meet the prescribed key behavior properties. It will help the developers design the software system with high reliability and functions satisfying the requirements.

At present, the formal methods in the aviation industries have been mainly applied in the process of software verification, and the schematic of software formal verification method is shown in Fig. 10.2. Honeywell Inc. have adopted the formal methods to verify 60% airborne software in the flight control system, thus improving 5 times of the software development efficiency and reducing the coding error to 1%. Besides, the system has also passed the certification of FAA.

At present, the formal method has been widely used in the airborne development process of aviation industries. Aimed at the utilization of this method in the aviation airborne software, the RTCA (Radio Technical Commission for Aeronautics) has proposed in DO-178B to use the formal

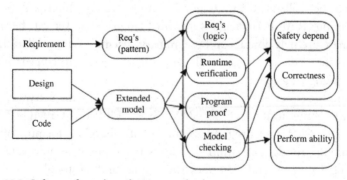

Figure 10.2 Software formal verification method.

method as an alternative compliance method of the software in airborne systems and equipment. And in the latest published DO-178C document, the RTCA further pubished DO-333 (Formal Methods Supplement to DO-178C and DO-278A) as the guideline of the formal method in the software life cycle.

10.4 FORMAL METHODS AND TOOLS USED IN THE SYSTEM DESIGN AND EVALUATION FIELD

10.4.1 SysML

In order to meet the actual needs of systematic engineering, a standard modeling language of systematic engineering on the basis of reusing and expansing the subset of unified modeling language UML2.0 is proposed by INCOSE and OMG, namely Systems Modeling Language (SysML) [16]. It is the expansion and customization of the UML so as to support the systematic modeling in systematic engineering on a large scale. The objective of SysML is to provide a standardized modeling language for systematic engineering, so as to analyze, describe, design, and verify the complex systems for improving the capability of interaction of systematic engineering information between different tools and helping build semantic connection of systems, software, and other engineering disciplines.

SysML supports the description, analysis, design, and verification of complex systems (including hardware, software, information, process, personnel and equipment, etc.) on a large scale. SysML is able to conduct modeling on various problems of systematic engineering in specific areas [17]. It can be used in different stages of systematic engineering, especially in that of detailed description and design. It is effective to use SysML to describe the requirement, systematic structure, functional behavior, and allocation. It eliminates the differences of various methods in the expression and terminology and avoids unnecessary confusion in symbolic representation and understanding. SysML is independent of any process and method of systematic engineering, but it supports the process and method. SysML emphasizes standard modeling language rather than standard process or method. SysML incorporates the advantages of systematic structure of object- and process-oriented methods, which can easily describe the connection and data exchange among systems. It further demonstrates that the application of SysML is inevitably related to the process. Different applications of systematic engineering require different processes. The development process proposed by the founder of SysML is an iterative and incremental process centered on system framework which is model driven [18].

It can be seen from Fig. 10.3 that SysML can be divided into three parts [19], namely, behavior diagram, structure diagram, and requirement diagram. Behavior diagram establishes a systematic behavioral model through the activity diagram, sequence diagram, state machine diagram, and use case diagram; structure diagram establishes systematic structural model through package diagram, block diagram, and parametric diagram; and requirement diagram establishes requirement model through requirement diagram.

SysML has been widely applied to various industries and be used in complex software systems at home and abroad to conduct modeling design so as to work out relevant products. The design of systematic integrated system and structure needs not only the standard language of system design—SysML, but also a systematic design standard which is universally recognized and authoritative. Therefore, Department of Defense (DoD) released "Department of Defense Architecture Framework, DoDAF" [20] version 1.0 in February, 2004, to be used for guiding the design of integrated systematic structure of defense command

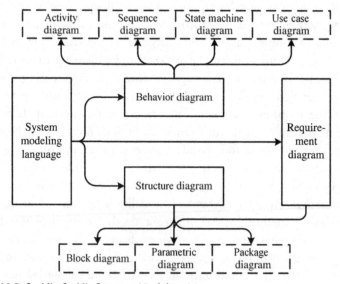

Figure 10.3 SysML. *SysML*, Systems Modeling Language.

and control system as well as business operation process. Subsequently, based on SysML, the design conception of DoDAF products is formed, and good application effects are achieved. Besides, Telelogic released TAU and Rhapsody products which support the SysML language. It incorporates the supporting environment of SysML modeling into these tools, which can be downward compatible with modeling languages such as UML and DoDAF.

10.4.2 AltaRica

AltaRica is a high-level modeling language initially dedicated to safety analysis of safety-critical system. AltaRica modeling language was developed in LaBRI in the 1990s [21]. A few years later, the AltaRica data-flow version was developed to handle industrial scale models. A number of processing tools have been developed for AltaRica, such as compilers to fault trees, compilers to Markov chains, generators of critical sequences, model-checkers, and stochastic simulators. AltaRica 3.0 is a major evolution of the language. Its underlying mathematical model— Guarded Transition Systems (GTSs)—makes it possible to design acausal components and to handle looped systems. The development of a complete set of freeware authoring and assessment tools is planned, so to make them available for a wide audience [22]. Fig. 10.4 presents the overview of the AltaRica 3.0 project.

AltaRica modeling can be divided into syntax and semantic levels. The form of expression at the syntax level is to model the system into a collection of Node possessing hierarchical structures. Each Node is characterized by various states, events, input/output, and data variables. The detailed behavior is expressed in a way similar to automata. AltaRica is a formal model of GTSs at the semantic level. The syntax model of AltaRica is transformed into the corresponding GTSs model through compilation and can generate the system fault tree model and further carry out the follow-up analysis work such as establishing corresponding temporal logic formula. After obtaining the fault tree model from AltaRica model, safety analyses based on fault tree can be further implemented by adopting the open-source tool XFTA on the basis of AltaRica. The output information includes typical minimal cut set, top/-basic event probability, and so on.

AltaRica Data-Flow is at the core of several Integrated Modeling and Simulation Environments: Cecilia OCAS (Dassault Aviation), Simfia

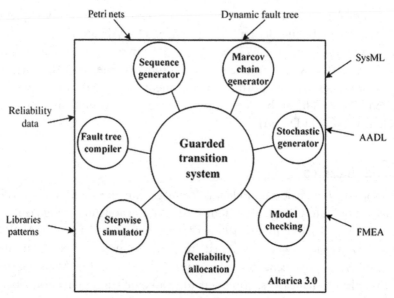

Figure 10.4 Overview of the AltaRica 3.0 project.

(EADS Apsys), and Safety Designer (Dassault Systèmes). Successful industrial experiments have been conducted by using AltaRica Data-Flow (e.g., certification of the flight control system of the aircraft Falcon 7X) [23]. In a word, AltaRica Data-Flow has reached an industrial maturity.

10.4.3 AADL

AADL [24] is a textual and graphical ADL that provides precise execution semantics for modeling the architecture of software systems and their target platform. It has been approved and published as an international standard by the International Society of SAE. A prototype of AADL was previously developed by Honeywell under US Government sponsorship (DARPA and others). This prototype, called MetaH, has been used extensively to validate the concepts in AADL. AADL is characterized by all the properties (e.g., composition, abstraction, reusability, configuration, heterogeneity, and analysis) that an ADL should provide. It is available for AADL to analyze the impact of different architecture choices (e.g., scheduling policy or redundancy scheme) on a system's properties. Therefore, AADL due to these characteristics, should be taken into full/careful consideration in the embedded safety-critical industry (e.g., Honeywell,

Rockwell Collins, Lockheed Martin, the European Space Agency, Astrium, Airbus, etc.) during the last years.

The AADL standard allows to describe both software architectures and execution platform architectures using one of its three complementary syntaxes:

- Textual: This is the reference syntax. It is more expressive than the graphical syntax and clearer than the XML syntax. It allows a complete description of an AADL model.

- Graphical: This syntax is complementary to the textual syntax, as it provides a global view of a system's architecture. However, the graphical syntax is not sufficient to describe complex and large systems.

- XML: This syntax is as expressive as the textual one and is designed to be a tool interchange format.

An AADL description consists of a set of component declarations. These declarations can be instantiated and connected to form a particular system architecture description. System descriptions in AADL allow a system designer to analyze the system schedulability, sizing, dependability, and other quality attributes. The AADL language has been designed to be extensible through annexes. The Error Model Annex is a standardized annex [25] that complements the description capabilities of the core AADL language by providing a textual syntax with precise semantics to be used for describing dependability-related characteristics in AADL architectural models (faults, failure modes, repair policies, error propagations, etc.). An AADL architectural model can be annotated with dependability-related information and the resulting annotated model can be used as an input to dependability analysis during different phases of the development cycle. A detailed guide for dependability modeling using AADL is available in Ref. [26].

In AADL, systems are modeled as hierarchical collections of interacting application components and a set of execution platform components. The application components are bound to the execution platform [27].

The AADL Error Model Annex supports the definition of reusable error models within libraries. Error models represent (stochastic) state machines that describe behaviors in terms of logic error states in the presence of faults, repair events, and error propagations. The user associates error models with application components, execution platform components, as well as the connections between them. When an error model is associated with a component, it is possible to customize it by setting component-specific values for the arrival rate or the occurrence probability for error events and error propagations declared in the error model.

10.4.4 Safety Cases

The safety case is a living set of documents which evolve over the lifetime of the system. In practice, the arguments of the safety case are contained in the safety case report, a document defining and describing the overall safety case, with references to a number of supporting documents [28] It is important that an adequate safety case is produced for a safety-related system in order to:

- ensure an adequate level of safety
- ensure that safety is maintained throughout the lifetime of the system
- minimize the licensing risk (being able to demonstrate safety to the regulators and assessors)
- minimize the commercial risk (ensuring implementation and maintenance costs are acceptable)

Safety Case is "A documented body of evidence that provides a convincing and valid argument that a system is adequately safe for a given application in a given environment." It includes the following aspects:

1. Make an explicit set of claims about the system
2. Provide a systematic structure for marshaling the evidence
3. Provide a set of safety arguments that link the claims to the evidence
4. Make clear the assumptions and judgments underlying the arguments
5. Provide different viewpoints and levels of detail

   Preliminary safety case element:

   - It establishes the system context (whether the safety case is for a complete system or a component within a system) and the phase of the project life cycle.
   - It also establishes safety requirements and attributes for the level of the design and interfaces to the system safety analysis.
   - It defines operational requirements and limits such as maintenance levels, repair time.

   Architectural safety case element:

   - It defines the system or subsystem architecture and makes trade-offs between the design of the system and the options for the safety case.
   - It defines the assumptions that need to be validated and evidence to be provided in the component safety cases.

- It also defines how the design addresses the preliminary operating and installation aspects for the safety case (e.g. via maintainability, modifiability, and usability attributes).
  Implementation safety case element:
- This safety case provides the justification that the design intent of the architectural safety case has been implemented and that the actual design features and development process followed demonstrate that the safety requirements are satisfied.
- Additional assumptions for operation and maintenance are identified and details are provided on how to meet the operational requirements.
  Operation and installation safety case element:
- This safety case adds details to the maintenance and support requirements identified in the implementation safety case.
- It defines any safety-related operational procedures identified in the preliminary safety case or Architectural safety case.
- For a COTS system, the safety case would include the safety justification of the specific configuration, and human factors—related issues such as staffing requirements and competence levels, training of operators and maintenance personnel, and facilities for long-term support.
- The safety case would also record and resolve any noncompliances with the original safety requirements.

10.4.5 NuSMV

NuSMV [29] developed by Carnegie Mellon and Trento has already been successfully applied to various safety analysis areas. It is applicable to technological transformation engineering, which is structurally sound, open and flexible, and close to industry standard in robustness.

The first version of NuSMV (herein after called the NuSMV1) basically implemented the BDD-based symbolic model checking. The new version of NuSMV (herein after called NuSMV2) inherits all the functionalities of the previous version, and extends them into several directions. The main novelty in NuSMV2 is the integration of model checking techniques based on propositional satisfiability (SAT) [30]. The SAT-based model checking is currently enjoying a substantial success in several industrial fields and opens up new research directions. BDD- and SAT-based model checking are often able to solve different classes of

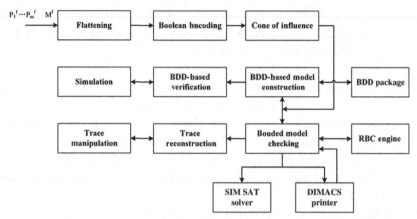

Figure 10.5 The internal structure of NuSMV2.

problems, and can therefore be regarded as complementary techniques. In order to integrate SAT- and BDD-based model checking, a major architectural redesign was carried out in NuSMV2, with the goal of making as many functionalities as possible independent of the actual model checking engine used. This allows the effective integration of the new SAT-based engine and opens up toward the implementation of other model checking procedures. A high-level description of the internal structure of NuSMV2 is given in Fig. 10.5

The most important feature of NuSMV is that it is easy to be modified and expanded as an open system, for the structure of NuSMV is formed and organized by blocks. In the NuSMV model, the system is expressed in a modularized hierarchical structure and the reusing of definition component is allowed. Each module can implement a set of function and communicate with other modules through a precisely defined interface. NuSMV is available in different areas, it is only necessary to replace or move certain input block. Based on the NuSMV input language developed for finite state system verification, main types of supporting data include the enumeration, Boolean, and capped array. Usually, a complete NuSMV model document consists of system model and system nature.

The FSAP/NuSMV-SA platform has been developed within the ESACS project [31], involving several research institutions and leading companies in the fields of avionics and aerospace. FSAP/NuSMV-SA aims to improve the development cycle of complex systems by providing a uniform environment that can be used both at design time and for safety assessment. The platform consists of two main components: Formal Safety Analysis Platform (FSAP), which provides a graphical front-end to the user; and NuSMV-SA based on the NuSMV model checker, which provides the safety assessment capabilities. FSAP/NuSMV-SA has been evaluated in collaboration with an industrial partner and tested on some industrial case studies.

10.4.6 PRISM

PRISM has been developed at the University of Birmingham. The core functionality of PRISM, namely, constructing a probabilistic model and then evaluating the results of one or more corresponding properties, is available to either a command-line or a graphical user interface. It provides direct support for the analysis on three types of probabilistic models: Discrete-Time Markov Chains (DTMCs), which made the behavior at each discrete time-step is modeled by a discrete probability distribution; Continuous-Time Markov Chains (CTMCs), which allow transitions to occur in real-time, with delays exponentially distributed; and Markov Decision Processes (MDPs), which combine discrete probabilities and nondeterminism for accurate modeling of concurrency. PRISM also provides indirect support for model checking probabilistic timed automata, including probability, nondeterminism, and real-time using clocks. For an overview of all four types of models and the techniques, it can be used to analyze them.

PRISM takes such a model description and constructs the corresponding probabilistic model, a process which includes determining the set of all reachable states of the model. Subsequently, the model can be analyzed by verifying that the system satisfies the properties specified in the temporal logic. PRISM uses the logic PCTL for properties of DTMCs and MDPs, and the logic CSL for properties of CTMCs.

In the algorithm, based on BDD and multiterminal BDD (MTBDD), PRISM incorporates the state symbol data structure and algorithm. Meanwhile, it includes a simulation engine of discrete event, providing support for similar/statistics model checking and achieving different analysis technologies such as quantitative abstraction refinement and symmetry reduction. Fig. 10.6 shows the PRISM system architecture.

PRISM [32] integrates completeness and powerful probability model experience. As an open-source probabilistic model checking tool, it can be used to conduct formal modeling on random or probabilistic behavior

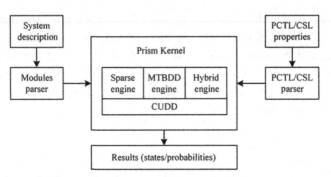

Figure 10.6 PRISM system architecture.

and be applicable to random systems, so it is widely applied in different areas, including communication, multimedia, random distribution algorithm, security protocol, biological system, etc.

10.4.7 SPIN

Simple Promela Interpreter (SPIN), developed by the Bell Labs, is a common tool used for formal verification. It supports the design and verification of distributed systems (e.g., data communicating protocol, distributed OS, database system, etc.), mainly demonstrating the correctness of process interaction. Processes in SPIN are considered to be asynchronous, and the explicit declaration is required for the synchronous processes. Through the implementation of analog systems and the generation of C program, SPIN can search the state space of the system. The nature in SPIN to be verified shall be expressed by linear temporal logic formula.

SPIN starts from the high-level model specification of a parallel system or distributed algorithm described by the system input language PROMELA (XSPIN is often used as the graphical front-end of SPIN). It can detect the syntax errors of the program and simulate the interaction of the system until the design is confirmed to have expected behaviors. Finally, SPIN will produce an optimized checking procedure written in C language which is compiled and then executed by the checker. If any counter-example of correctness statement is found, then it shall be fed back to the interactive simulator to help users find out the cause of the error. Fig. 10.7 describes the checking process of SPIN.

The good algorithm design and extraordinary detectability of SPIN win the recognition of the world's best professional Association for

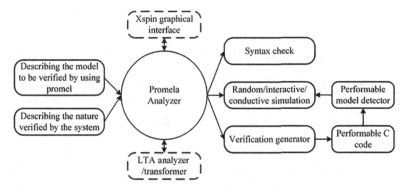

Figure 10.7 Basic structure of SPIN. *SPIN*, Simple Promela Interpreter.

Computer Machinery (ACM). In 2001, Holzmann was awarded the famous "System Software Award" [33]. NASA used SPIN to detect errors of Mars explorer occurred as early as in 1996, and it was found that some errors could be corrected before the launch. Since then, SPIN has been used to detect the Saturn rocket control software and some procedures applied to the outer space. Lucent company also discovers the advantages of SPIN, and PathStar Access Server is the first Lucent product which benefits from Holzmann (the developer of SPIN). Holzmann uses SPIN to detect the code of a new version of 5ESS Switch. In addition, SPIN has been successfully applied in the security protocol verification, data communication, software verification, optimal programming, and other fields.

10.5 A CASE STUDY OF THE FORMAL METHOD USED IN THE SAFETY ASSESSMENT

In this chapter, the safety analysis is conducted in a formal method on the power system that adopts the system architecture shown in Fig. 5.10.

The first step is to collect the system information, including the system architecture, the relationship characteristics between the input and output of the module.

The second step is to determine the failure conditions to be analyzed by the same way adopted in the case of Chapter 6, Common Cause Analysis, namely, "Loss of the L DC Bus, R DC Bus, and DC ESS Transfer Bus."

The third step is to set up the failure expansion model of each component in the power system, including the input and output relations of each module, the failure event of the module, and the transformation relationship between the possible states of the module. The failure expansion model, established for each component of the power system, is shown in Figs. 10.8−10.14.

The fourth step is to establish the failure logic according to the failure condition. For the failure condition of the "Loss of the L DC Bus, R DC Bus, and DC ESS Transfer Bus" analyzed in this case, the failure logic is as follows:

L_DC_Load = False & R_DC_Load = False & L_DC_ESS_Bus = False & R_DC_ ESS_Bus = False.

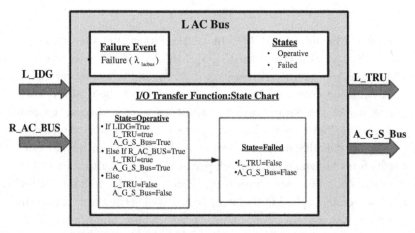

Figure 10.8 L AC BUS component block.

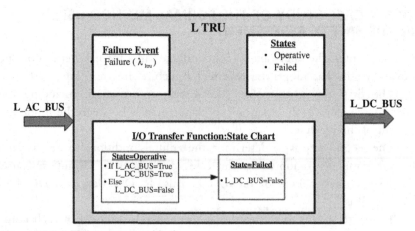

Figure 10.9 L TRU component block.

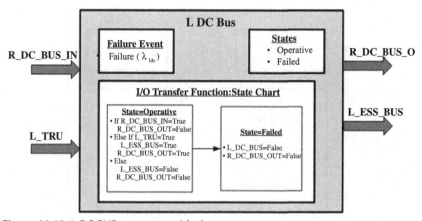

Figure 10.10 L DC BUS component block.

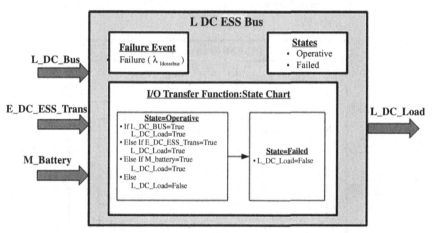

Figure 10.11 L DC ESS BUS component block.

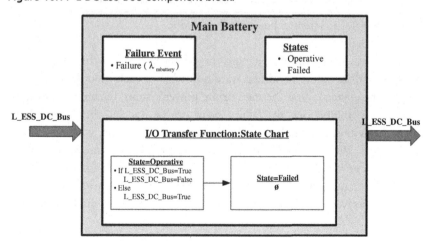

Figure 10.12 Main battery component block.

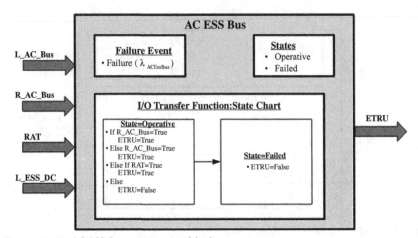

Figure 10.13 AC ESS Bus component block.

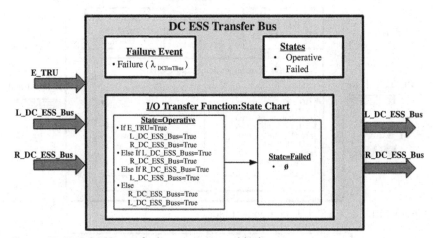

Figure 10.14 DC ESS transfer bus component block.

The fifth step, through injecting the failure event into the system, is to make use of the failure condition transition table of each module to judge its output, thus determining the state of the power system. In the case of multiple failure combinations, the failure event can be injected simultaneously to judge the system state. Followed the fault propagation logic, when the L DC Bus, R DC Bus, and DC ESS Transfer Bus are all in failure states, then a cut set can be obtained. The analysis results are shown in Fig. 10.15. As we can see, the cut-set result derived by the MBSE method is just the same as the one derived by FTA. For a system composed of multiple system

Trace number	Failure Event Injected	L AC Bus				L TRU		L DC Bus			...	Main Battery	
		In		Out		In	Out	In		Out		In	out
		L IDG	R AC BUS	L TRU	A G S Bus	L AC Bus	L DC Bus	L TRU	R DC Bus	L ESS DC Bus		L ESS DC Bus	L ESS DC Bus
1	L AC Bus Failure	T	T	T	T	T	T	T	T	T		T	F
		T	F	F	F	F	F	F	T	T		T	F
		F	T	T	F	F	F	F	F	F		F	T
		F	F	F	T	T	T	T	F	T		F	T
2													
3													
N	L IDG Failure & R IDG Failure & RAT Failure	F	F	F	F	F	F	F	F	F	...	F	T

components, the traversal way causes too much work, while appropriate modeling tools can be used to complete the analysis.

REFERENCES

[1] Davies J, Schneider S. A brief history of timed CSP. Theor. Comput. Sci. 1995;138:243—71.

[2] Jones CB. Systematic software development using VDM[M]. 2nd ed. Prentice Hall: Englewood Cliffs, NJ; 1990.

[3] Peterson JL. Petri nets. ACM Comput Surveys 1997;9(3):223—52.

[4] Brauer W, Reisig W, Rozenberg G, et al. Petri nets: central model and their properties. In: Lecture Notes in Computer Science(LNCS). New York: Springer-Verlag; 1987.p. 87—90.

[5] Bertoli P, Bozzano M, Cimatti A. A symbolic model checking framework for safety analysis, diagnosis, and synthesis. In: Model checking and artificial intelligence. Springer Berlin Heidelberg; 2007.p. 1-18.

[6] Cubbin C. ISAAC, a framework for integrated safety analysis of functional, geometrical and human aspects. In: Proc.ERTS; 2006.

[7] More Integrated Systems Safety Assessment. http://www.missa-fp7.eu/.

[8] Bozzano M, Villafiorita A, Akerlund O, et al. ESACS: an integrated methodology for design and safety analysis of complex systems. In: Proc. ESREL 2003; 2003.

[9] Bernard R, Aubert JJ, Bieber P, et al. Experiments in model based safety analysis: flight controls. Dependable Control Discrete Syst 2007;1(1):43—8.

[10] Bozzano M, Villafiorita A. The FSAP/NuSMV-SA safety analysis platform. Int J Softw Tools Technol Transf 2007;9(1):5—24.

[11] Joshi A, Whalen M, Heimdahl MPE. Model-based safety analysis final report. NASA Contractor Report NASA/CR-2006-213953; 2006.

[12] Meenakshi B, Das Barman K, Babu KG., et al. Formal safety analysis of mode transitions in aircraft flight control system. In: Digital Avionics Systems Conference; 2007. p. 2.C.1-1—2.C.1-11.

[13] Elmqvist J, Nadjm-Tehrani S. Formal support for quantitative analysis of residual risks in safety-critical systems. In: High assurance systems engineering symposium, 2008. HASE 2008. 11th IEEE. IEEE; 2008. p. 154—64.

[14] Lindsay PA, Winter K, Yatapanage N. Safety assessment using behavior trees and model checking. In: Third IEEE international conference on software engineering and formal methods (SEFM'05). IEEE; 2010. p. 181—90.

[15] Gomes A, Mota A, Sampaio A, et al. Systematic model-based safety assessment via probabilistic model checking. Lect Notes Comput Sci 2010;6415:625—39.

[16] Hause M, Thom F. Modeling high level requirements in UML/SysML. In: INCOSE Symposium 2005, Rochester, US; 2005.

[17] Friedenthal S, Moore A, Steiner R. A practical guide to SysML: the systems modeling language. Amsterdam: Elsevier; 2011.

[18] Weilkiens T. Systems engineering with SysML/UML: modeling, analysis, design. Burlington, MA: Morgan Kaufmann; 2011.

[19] Jiang CY, Wang WP, Li Q. SysML: a new systems modeling language. J Syst Simul 2006;18(6):1483—7.

[20] DoD Architecture Framework working Group. DoD architecture framework version 1.0, volume II: product descriptions. The United States: Department of Defense; 2004.

[21] Arnold A, Griffault A, Point G, et al. The altarica language and its semantics. Fundament Inform 2000;34:109—24.

[22] Rauzy A. Guarded transition systems: a new states/ events formalism for reliability studies. Proc IMechE, Part O: J Risk Reliab 2008;222(4):495−505.

[23] Bernard R, Aubert JJ, Bieber P, et al. Experiments in model-based safety analysis: flight controls. In: Proceedings of IFAC workshop on dependable control of discrete systems, Cachan.

[24] SAE-AS5506. SAE Architecture Analysis and Design Language (AADL). International Society of Automotive Engineers; November 2004.

[25] SAE-AS5506/1. SAE Architecture Analysis and Design Language (AADL) Annex Volume 1, Annex E: Error Model Annex. International Society of Automotive Engineers; June 2006b.

[26] Feiler PH, Rugina AE. N CMU/SEI-2007-TN-043 Dependability modeling with the architecture analysis and design language (AADL). Carnegie Mellon Software Engineering Institute; 2007

[27] Ana ER. Dependability modeling and evaluation—from AADL to stochastic Petri nets [Ph. D. Thesis]. Toulouse: LAAS, 2008.

[28] MoD. Defence Standard 00−56. Safety management requirements for defence systems, part 1: requirements, Issue 4; 2007.

[29] NuSMV: a new symbolic model checker. http://nusmv.fbk.eu/.

[30] Biere A, Cimatti A, Clarke EM, Zhu Y. Symbolic model checking without BDDs. In Proc. of the fifth international conference on tools and algorithms for the constructionand analysis of systems (TACAS'99); 1999.

[31] Bozzano M, Villafiorita A, Åkerlund O, et al. ESACS: an integrated methodology for design and safety analysis of complex systems. In: Proceedings of the European Safety and Reliability Conference (ESREL 2003); 2003. p. 237−45.

[32] Kwiatkowska M, Norman G, Parker D. PRISM 2.0: a tool for probabilistic model checking. In: Quantitative evaluation of systems, international conference on. IEEE Computer Society; 2004. p. 322−3.

[33] Clarke EM, Wing JM. Formal methods: state of the art and future directions. In: ACM computing surveys; 1996. p. 626−43.

INDEX

Note: Page numbers followed by "*f*" and "*t*" refer to figures and tables, respectively.

Printed in the United States
By Bookmasters